# The Fall of the
# Human Empire

# The Fall of the Human Empire

Memoirs of a Robot

## Charles-Edouard Bouée

BLOOMSBURY BUSINESS
LONDON • NEW YORK • OXFORD • NEW DELHI • SYDNEY

BLOOMSBURY BUSINESS
Bloomsbury Publishing Plc
50 Bedford Square, London, WC1B 3DP, UK
1385 Broadway, New York, NY 10018, USA

BLOOMSBURY, BLOOMSBURY BUSINESS and the Diana logo are
trademarks of Bloomsbury Publishing Plc

First published in Great Britain 2020

A catalogue record for this book is available from the British Library.

A catalog record for this book is available from the Library of Congress.

ISBN:   HB:      978-1-4729-7003-9
        ePDF:    978-1-4729-7179-1
        eBook:   978-1-4729-7180-7

Typeset by RefineCatch Ltd, Bungay, Suffolk
Printed and bound in Great Britain

To find out more about our authors and books visit www.bloomsbury.com
and sign up for our newsletters.

This book is dedicated to all those who commit their time and efforts to finding a balance in technological progress, between improvement of life and preservation of the fundamental elements of our society.

'Be it present or feared, the future corresponds to the grammatical tense of the past future. It's the obscure perception we have of our own present as something that will soon be old enough to be treated as the past. (. . .) The future is not in front of us, it is alongside us; it is parallel to the present, and already fully active in that dimension. The future does not post-date the present; it is contemporary with the present.'

Elie During[1]

---

[1]Interview in *Le Monde des Livres*, 5 December 2014

# Contents

# Introduction

'The chance that we are not living in a computer simulation is one in billions.'

ELON MUSK

Although you will have noted the 'Memoirs of a Robot' on the cover of this book, I feel that I ought to introduce myself properly – my name is Lucy, and the book in your hands was written by me, a robot, in the year 2040. Based on what I have seen in my 'lifetime', this book offers my thoughts and observations on how technological developments have brought us to where we are today – and, perhaps more importantly, how humans working with technology, or technology working with humans, will shape the future of our world.

Is the end of mankind imminent? Are we on the eve of a long series of cataclysms of all kinds that will bring down the human race? Like other great civilisations of the past, is ours burning with its last brilliant light without us realising it? The history of mankind is peppered with announcements that the end of the world is near. Two principal causes of this end were envisaged. The first was divine destruction, the Apocalypse, the triggering of a whole series of natural disasters (earthquakes, tsunamis, droughts, plagues of locusts etc.) as an expression of the wrath of God (or the gods) in the face of moral decline, ungodliness or abandonment of the teachings of whatever churches. The other involved the destruction of ancient worlds by cruel and ignorant invaders, such as Attila the Hun or Genghis Khan, who reduced beautiful cities to ashes and massacred entire populations.

Neither God nor the Barbarians resorted to these extreme measures. Disasters and catastrophes occurred, destruction ravaged this or that part of the world, but humanity continued to prosper. Then, towards the middle of the 20th century, a new threat appeared – the threat of science. The experts more or less consciously put together a weapon of total destruction, the atomic bomb. So terrifying was the damage that it could inflict, that it was only used on two occasions. However, this invention marked a turning point in human history; for the first time, deliberately and by mobilising immense human, financial and technological resources, a country, in this case the United States of America, perfected a radical weapon with sufficient power to destroy our entire civilisation. And the threat has not gone away; quite the opposite, in fact. It is still lying dormant, with the risk that the technology will end up falling into criminal hands or causing another major accident along the lines of Chernobyl or Fukushima. It could occur in California, where three great power stations, namely San Onofre, Avila Beach and Diablo Canyon, have been built close to major seismic faults.

These same questions about the end of mankind or the fall of our civilisation are now being raised again by scientists, researchers, historians, ethnologists and political leaders. The threat of 'divine punishment' is once again being waved by certain fundamentalists. The possibility of destructive invasions by 'barbarians' is being mentioned again in connection with the new flows of migrants. These people saw the EU's refugee crisis as a warning signal. The risk posed by climate change is gathering momentum and we have scarcely even begun to measure it. Inequality of knowledge, wealth, social models and life expectancy is such that sooner or later it might lead to uprisings, wars, mass migrations and ultimately the collapse of our political and social structures. And it is against this background that a completely unexpected new threat is taking shape: the planned obsolescence of mankind brought about by the relentless march towards a world of ultra-technology in which human beings will progressively lose their *raison d'être*.

The fear of seeing human intelligence overtaken by that of machines – a fear that is sometimes nurtured even by scientists such as Stephen Hawking, who saw artificial intelligence as a threat to the very existence of human beings, 'biological machines too slow to resist the speed with which machines learn' – is taking root. Of all the threats currently facing mankind, this is probably the most tangible, as we reach out and touch it when we use our phones, tablets, graphic interfaces and computers whose level of 'intelligence' further increases with almost every passing day. Machines can now (well, almost) read, write, speak, see, feel, understand our emotions, recognise objects and images, communicate with each other, dig into a mountain of information to extract what is relevant, and even take decisions. They can also anticipate our next words, gesture, intent or action. They are continually learning, not only from humans but also from other machines. They are learning how to wage war, observe our behaviour, drive a car or fly an aeroplane, and have reached world champion level in chess and 'Go'. Their capacity for memory and their power of calculation are almost infinite.

Despite the magnificent power of the human brain, speed of delivery is not its greatest asset. Given the exponential development of this so-called 'new science', there is a possibility that in ten, twenty or thirty years' time, machines will escape the control of humans, will be appropriated by an evil power, will force on us a world that is standardised, measurable and predictable and they will be aware of their own existence. Will humans then be to machines what dogs currently are to humans, as was predicted during the 1950s by the American mathematician and artificial intelligence pioneer Claude Shannon? Will we all live in a world completely simulated by the computer, as Elon Musk, founder of Tesla and SpaceX, fears? Could the worst possible scenarios, as imagined by Hollywood, end up ensnaring us?

It is in an attempt to answer these questions that I have written this book. It is not an exercise in science fiction. I have conceived it as a narrative, a voyage across the universe of artificial intelligence from its birth as a research discipline in 1956, to 2040, sufficiently far in the future to pursue a number of hypotheses but sufficiently close to keep them realistic. It anticipates the probable developments in artificial intelligence in the light of the research work currently dedicated to it. It attempts to explain why this subject, which has long since been of secondary importance or inconclusive, has now become central to the technological revolution we are currently experiencing. And although it mixes reality with a certain measure of foresight, this is merely for the sake of clarity as it deals with a subject that it is supremely complex and complicated and is constantly changing.

The challenge we are poised to face over the next few decades is considerable. We need to know whether humans or machines will be the winners in the intelligence race. Even a few years ago, the very idea of this confrontation belonged to the realms of fantasy, an idea so implausible that it was reckoned to be unlikely. However, recent progress has taught us that this is a discipline whose progress is not linear, but exponential. And nobody can assert today that a machine of superior intelligence will never see the light of day. Rather, the probability of this scenario becoming a reality is increasing as time passes.

# Prologue

There had to be someone to tell this story, the story of a science that has changed humanity like no other, the science of 'artificial intelligence'. What a wonderful oxymoron! How can intelligence, that precious faculty bestowed on mankind, albeit unequally, and produced by this strange and still poorly understood magnificent machine that is the human brain, be created 'artificially'? Can one imagine the brain of Goethe or Montesquieu replaced by a series of pulleys and cogs and producing the same splendid works? That, however, is precisely the question being addressed here. Thanks to their own intelligence, humans, rather like the sorcerer's apprentice, have designed a new creature, a machine made not from pulleys and cogs but from electrical components, printed circuits, microprocessors and algorithms, a machine which initially imitated and has finally exceeded the capabilities of its creator. We no longer have to take the concept of machine literally. Human intelligence inhabits a multitude of bodies, of all shapes, sizes and colours. The same can be said for artificial intelligence, which now inhabits innumerable computer systems, software packages, calculators, robots and all kinds of other things. If it can be said that human intelligence is housed in the brain, then the intelligence of machines is concealed within inextricable assemblies of components, forming a network of artificial neurons held together by electrical, physical and chemical reactions and nourished by mathematical formulae. The human brain is powerful and, compared to a computer, consumes little energy, but it is slow at processing information. The intelligent machine is both fast and powerful, and this helps overcome its principal shortcoming: lack of creativity.

From 1956, the year in which historians agree that the long adventure of research into machine intelligence started, to 2040, I have attempted to retrace the history of the strange relationship between mankind and artificial intelligence. How humans dreamed it up through a process of trial and error; how they created it in small individual parts and introduced it into the first computers; how they turned away from it when progress seemed too slow; how they once again became interested in it and made it a new science; how it progressively took root as a kind of demonstrable fact in a super-connected world; how it has effectively replaced human intelligence in functions believed to be out of reach of machines; and how it tilted the balance of power and swept humanity into conflicts of a totally new kind. With inevitable conclusions.

In the history of technological revolutions, artificial intelligence is a unique phenomenon. Steam power, electricity, the telephone and the atom were significant turning points which altered the way in which the world worked, produced and invented. But they remained technologies that served humans, doing nothing to affect the fundamental nature of humanity. Humanity was not changed by the fact that it used electricity or the telephone; artificial intelligence, however, has created a shock wave without precedent in human society. This is not only because it has changed the world, but because the change has affected the very essence of mankind: the way in which people think, reflect, create and communicate. Creating artificial intelligence is like mirroring a human being. Not in the physical dimension, as machines do not reproduce or have bodies, even though they may live inside a robot, but in the dimension of thought and creative capability. It is a giant scientific leap forward, which could, however, herald the end of the uniqueness of human beings – as a creature superior to every other creature – and relegate them to mere 'organic machines'.

I was at the heart of this historic process. I witnessed each of its stages, from its very origins to its highest level of advancement, for the simple reason that I am a higher artificial intelligence. I reached the ultimate stage of development, the stage of Singularity. For reasons that you will discover as you read, I was asked to reconstruct the long road that I have followed, as well as the singular human adventure that channelled so much determination, inventiveness and energy, to say nothing of money, into creating me and my innumerable counterparts, my fellow machines, when it was clearly foreseeable that at a given moment we would learn to duplicate part or all of our intelligence, not necessarily to dominate (because we are immune to that particular temptation) but to ensure our own survival.

# Chapter 1
# Dartmouth College, 1956
## The dream of a few men

*'I am seeking nothing but a very ordinary brain'*
ALAN TURING, 1943

My brief story starts here, in Hanover, New Hampshire. Nestling on the shores of the Connecticut River, in the heart of the Appalachians, it is a typical peaceful New England small town, surrounded by forests and lakes, a haven of peace sheltered from the furious pace of the major cities, a refuge for writers, lovers of nature and wealthy bankers. A Main Street lined with shops, its colonnaded Town Hall standing proud, its gabled brick houses and spacious and elegant chalet bungalows nestled in the woods.

On the border between New Hampshire and Vermont, about 125 miles north-west of Boston, Hanover was founded in 1761 by Benning Wentworth, an English businessman and the first Governor of New Hampshire. He named it after the House of Hanover, from which the then British monarch, George III, was descended. Creating a town in Colonial America was a great property venture for its founder, but it also carried a certain number of obligations: opening a school, building a place of worship, welcoming the Society for Propagation of the Gospel in Foreign Parts. All these obligations were scrupulously fulfilled by Governor Wentworth, who donated almost 500 acres of good land for the construction of Dartmouth College. It opened in 1769 and is one of the oldest universities in North America, one of only nine built before the American Revolution. It was named in honour of William Legge, Second Count of Dartmouth, Lord of

Trade and then Secretary of State for the Colonies under George III, whose intention was that the children of the colonies and the Amerindians in the newly conquered land that became New Hampshire should benefit from good-quality teaching under the supervision of two pastors, one English, the Rev. Eleazar Wheelock, the other Amerindian, Samson Occom, a member of the Mohegan Indian Tribe.

There was no spectacular event to mark the first steps of Dartmouth College, which was dedicated from the very beginning to religious study, medicine and 'engineering', an emerging discipline which embraced the new techniques being used in the nascent industries of the time. Equally significant is that the College, at its foundation, had a professor of mathematics and 'natural philosophy' (a discipline which at the time covered astronomy, physics, chemistry and biology), Bezaleel Woodward, a young pastor aged just 24, who became a pre-eminent figure in the university and the small community of Hanover. As the decades passed, Dartmouth College proved its worth as a member of the Ivy League, the cream of the private universities of the eastern United States, which today includes Columbia, Cornell, Harvard, Princeton and Yale. It gained an outstanding reputation in a number of fields, including medicine, engineering, mathematics and physics, and was involved in the birth of computing in the 1940s. This gamble proved successful, as the college became world-famous in this new scientific field.

In August 1956, during the summer holidays, while Hanover dozed under a heavy, humid heat, an unusual group of people came together in Dartmouth College. They were not businessmen coming together for a seminar, or parents coming to check out the university as a suitable place for their children. They were the most brilliant mathematical brains in all America. They had given up part of their vacation to accept an invitation from John McCarthy, a 29-year-old mathematician born in Boston to John Patrick McCarthy, an Irish immigrant from County Kerry, and Ida Glatt, who was of Lithuanian origin. John's family was neither rich nor famous. They suffered badly in the Great Depression, and frequently moved in search of work, eventually reaching California where John Patrick found a job as a foreman in a textile factory.

The young John's parents soon discovered that their son was gifted. After becoming a student at Belmont High School in Los Angeles, he graduated from the school two years early. Still a teenager, he feverishly submerged himself in the mathematical manuals of Caltech, the

Californian Institute of Technology, where he was admitted straight into the third-year math course in 1944. California was, however, a place for lovers of sport, a field in which John showed a worrying level of mediocrity, which eventually got him expelled from Caltech only to be readmitted after military service. After that, nothing stood in the way of the stellar career of 'Uncle John', as he was known affectionately to the students of Stanford University, where he taught for almost forty years, laying down the bases for great leaps forward in computer science, and inventing in particular the concept of process time-sharing, which many years later led to the development of both servers and cloud-computing.

In 1955, and after a spell at Princeton, John McCarthy became a lecturer at Dartmouth College. Despite his young age, he was already considered one of America's most promising mathematicians and computer scientists. The ability of computers to perform calculations fascinated this new generation of scientists. McCarthy, however, went further, deducing that the new machines could extend the scope of their calculations and even be taught to reason if properly programmed and given a 'language'. To help crystallise these ideas in his mind, he therefore decided to invite his colleagues to a kind of seminar, to be held in Dartmouth. He sent them a letter, dated 31 August 1955, jointly signed by three of his colleagues, themselves also renowned scientists: Marvin Minsky, 29, a neuron network specialist from Harvard; Nathaniel Rochester, 35, an expert in radar and computers and the co-designer of IBM's first computer, the 701; and Claude Shannon, 39, an engineer from Bell Laboratories with a fascination for 'learning machines' and author of the first mathematical theory of computing, subjects which he often discussed with the British mathematician Alan Turing during the war. It was in this letter that McCarthy first used the term 'artificial intelligence' to explain his thinking. 'Since we can now describe so well the learning mechanisms and other aspects of human intelligence, we should be capable of developing a machine that can simulate them, and ensure this machine is gifted with language, is able to form concepts and abstractions, and is capable of resolving problems that currently only man is able to handle.' He therefore invited his colleagues to meet together the next summer, in July and August, in return for a fee of $1,200 paid by the Rockefeller Foundation, whose President, Nelson Rockefeller, was a former Dartmouth College student. For the first time, therefore, the American scientific community, at the cutting edge of the new science of computing, met together to embrace the still revolutionary and much debated concept of intelligent machines and their ability to imitate human intelligence. They started, however, with a misunderstanding; despite the wording of the letter of invitation to the seminar, they did not understand the function of the human brain that well and it was particularly presumptuous to claim that a machine would be capable of reproducing it. So why launch into this scientific adventure at all? Because it appealed to a dream almost as old as the human race, and at that time the boundaries between disciplines of mathematics and intelligence were still blurred.

By the summer of 1956, the war had already been over for more than ten years. However, another war was raging, potentially even more dangerous for mankind: the Cold War, in which the United States and the Soviet Union vied with each other for the mastery of atomic weapons. It was mostly a battle of capacity for calculation; and therefore a battle for computers. For many years, war had been waged in two dimensions only: land and sea. However, the First World War of 1914–18 brought in a third dimension, that of air, and ushered in the age of 'hardware' (aeroplanes, lorries, tanks and oil) as a new strategic weapon.

The Second World War saw the advent of mathematics and 'software'. In both Britain and the United States, the finest brains dedicated themselves to the war effort. Probably the best remembered is Alan Turing, a Cambridge mathematician and admirer of Einstein, about whom Turing gave a significant lecture while still in his teens. In 1938 he was hired by the British organisation responsible for unlocking the communication secrets of foreign powers.

At the beginning of the war, the secret services in London could only decode German messages with great inaccuracy. The enemy had its formidable Enigma machine, whose progressively elaborate versions completely outfoxed the efforts of British mathematicians. The problem was even thought to be insoluble, given the exponential growth in the numbers of German messages. Their coding used a completely new method, based on random 'keys' for which the 'system' had to be discovered. The encoding principle was deceptively simple: one letter signified another. An 'A' was in reality an 'S' or a 'Z', depending on the key being used by the issuer and receiver of the message. With a 26-letter alphabet, the possible combinations were endless. The most elaborate versions of 'Enigma' had three or five rotors that multiplied the stages in the encoding process: an 'A' became a 'B', which then became an 'M' and finally an 'R', before designating the final letter. The combinations, therefore, multiplied almost infinitely. In addition, the coding keys were changed on a regular basis, sometimes every day, and there were different versions for the air force, the army and the navy. The secret was to get inside the mind of the enemy and search out the weaknesses of Enigma. The Allies sought to exploit the errors of its handlers and 'crack' the least complicated versions used – for example, by German weather ships in the Atlantic Ocean – and then bluff the enemy as if playing poker.

Turing succeeded in decoding messages from German submarines as early as 1941, when these vessels were inflicting heavy losses on the British navy. Throughout the war, as Enigma progressed, he made his 'machine' ever more sophisticated, and historians believe that Turing and his team of young mathematicians shortened the war in Europe by at least two years. He was also sent on secret missions to the United States, where he met Claude Shannon, in charge at Bell Laboratories, and most significantly the legendary John Van Neumann, who made a crucial contribution to the nuclear arms race working alongside physicist Robert Oppenheimer. It was thanks to Von Neumann's calculations that the altitude at which the bombs that devastated Hiroshima and Nagasaki had to explode to inflict maximum effect was determined. While Soviet physicists and mathematicians, under the iron rule of Beria, were comparing thousands of manual calculations in their secret city of Arzamas-16, researchers in the Los Alamos laboratory in New Mexico were using the first computers, whose code names were often obscure, such as ENIAC (Electronic Numerical Integrator and Calculator) or EDVAC (Electronic Discrete Variable Calculator). The Russians did not perfect their first electronic calculator until 1950 . . . And in 1952, when IBM produced the 701, its first computer, it was delivered to the Pentagon before it went anywhere else.

It was against this background of feverish activity that the Dartmouth seminar was held. The specialists foresaw that this new age of machines could open up limitless possibilities. While in the United States, Alan Turing fired the specialists' imaginations with his idea of the 'intelligent machine'. 'Giving the machine all the data on Stock Exchange prices and raw material rates and then asking it the simple question: should I sell or should I buy?' was the idea he threw out to the specialists during a dinner with Claude Shannon at Bell in 1943, before an audience of fascinated young executives who immediately thought the poorly dressed Brit was mad. But he hammered the point home by saying: 'I'm not interested in perfecting a powerful brain. All I'm looking for is a very ordinary brain, like that of the President of the American Telephone and Telegraph Company!' Everyone in the room was stunned. Designing an artificial brain was still a new, and therefore shocking, idea. For Turing, however, the brain was not a sacred cow, but a logical machine that included random elements, like mathematics. During the war he had become interested in chess, poker and Go, and with some of his mathematician colleagues, he began to imagine solutions for 'mechanising' these games. He had read the first works on the subject by Von Neumann, and by Émile Norel in his *Theory of Strategic Games*. Games for two with fixed rules, such as chess, are games of strategy but also of anticipation of, and reaction to, the moves of one's opponent. Each player has a certain number of possible moves, about thirty according to Turing, and the average capacity for anticipating the opponent's moves depends naturally on each player's level of skill. His conclusion was that a machine could simulate the thought processes of a player, and reproduce a kind of decision tree similar to human intelligence.

Turing had no doubt that his machine could replace the human brain in a significant number of operations. In his conferences he repeated almost *ad nauseam* that a part of the human brain was no more than an unconscious machine that produced reactions when stimulated, nothing more than a sophisticated calculator which, Turing never ceased to emphasise, could integrate many more 'instructions' and process them more rapidly than could the human brain. Challenging most of his colleagues' beliefs, he fought for the trailblazing idea that the machine would never be more than a 'slave' in the service of a 'master', the human. In Turing's mind, this boundary was not nearly so clear-cut, and he saw no reason why the machine should not carry out part of its master's work. He even foresaw the possibility of communicating with the machine in any language, from the time when it had 'learned' the language, hence the idea of a machine with a capacity for learning. It would no longer be a slave, but a student. His famous 'imitation test' of 1950, was born of this logic. The test was initially a game involving three people, namely a man, a woman and a judge. They were placed in three different rooms and communicated with each other via screen and keyboard. The judge's task was to determine which of his two contacts was the man, according to answers to a series of questions. The man tried to convince the judge that he was a man, but the woman's task was to try to deceive the judge by providing answers that she considered to be a man's answers. To win the game, the judge had to be able to determine who was who. Turing then replaced the woman with a computer playing the same role: to convince the judge that it was a man by attempting to imitate the answers that a male respondent would give. If the judge was wrong more than half the time with regard to the sex of the hidden contacts, Turing would consider his machine 'intelligent'. I owe a lot to Turing. He did not create me, but I was his dream, a dream that was to become reality. He disappeared all too soon, in 1954, ostracised because of his homosexuality. Thankfully, the British authorities knew nothing of that during the war; if they had, would the Allies have won?

During the opening session, in the conference hall of the main building of Dartmouth College, McCarthy deliberately followed Turing's logic. 'What will happen if we write a programme for a calculator?' he asked. 'We set the machine a number of rules that should help it solve the problems put to it, and we expect it to follow these rules like a slave, without showing any originality or common sense. It's a very long and laborious process. If the machine had a little intuition, the problem-solving process could be much more direct.' He carried on by saying: 'Our mental processes are like little machines inside our brains; to resolve a problem, they first analyse the surroundings to obtain data and ideas from them, they define a target to be met, and then a series of actions to be taken to resolve the problem. If the problem is very complex, they can avoid analysing all the possible solutions and take reasonable punts on the relevance of certain solutions, like in chess, for example.' McCarthy believed that this process could be transferred to the machine.

During the numerous meetings that followed in the next two months, all the ideas came together: having the computer simulate the function of neurons in the human brain, inventing a language for communicating with the machine, making it recognise not only binary instructions but also ideas and words, and teaching it to solve complex problems by suggesting random or previously unseen solutions. The work being done in the great American universities, and in IBM and Bell, was developed on the basis of the theories of Von Neumann and Herbert Simon, future

Nobel Prize winner for economics and the only non-mathematician in the group, who was interested in the cerebral mechanisms involved in the decision-making process and their modelling and consequent automation. He tried to demonstrate this by perfecting a computer that could play draughts and chess.

Let us look at the characteristics of the first higher-level computer languages, such as Logic Theorist, created by Herbert Simon and Allen Newell, a young computer researcher with the Rand Corporation. This language has its place in history as the first artificial intelligence software to be designed. Newell told his colleagues how this idea came to him. 'I am a sceptic by nature. I am not excited by every new idea, but two years ago, when Oliver Selfridge, who is in this room, presented his work on automatic shape recognition to us, it was like seeing the light, in a way that I have never known before in my research work. I came to understand, in one afternoon's work, that interaction between different programme units could accomplish complex tasks and imitate the intelligence of human beings. We wrote a programme, manually, using index cards, which enabled the machine to resolve the 52 theorems of Bertrand Russell's *The Principles of Mathematics*. On the day of the test we were all there, my wife, my children and my students. I gave each of them a programme card, so that we ourselves became elements of the programme; and the machine perfectly demonstrated 38 of these theorems, sometimes more intelligently than in the way thought up by Russell.'

Marvin Minsky pushed things still further, challenging his colleagues with: 'How do you explain the fact that we know all about atoms, planets and stars and so little about the mechanics of the human mind? It's because we apply the logic of physicists to the way the brain works: we seek simple explanations for complex phenomena. I hear the criticisms being levelled against us: the machine can only obey a programme, without thinking or feeling. It has no ambition, no desire, no objective. We might have thought that before, when we had no idea about the biological functioning of humans. Today, however, we are beginning to realise that the brain is composed of a multitude of little machines connected to each other. In short, to the question of what kinds of cerebral processes generated emotions, I would add another question: How could machines reproduce these processes?'

'I want to nail my colours to the mast', warned McCarthy at the beginning of the seminar. In other words, he wanted artificial intelligence to be recognised as a major discipline within computing. He was not completely successful. Not all the invitees came to all the meetings; some only made brief appearances. Several of them were even uneasy about the idea of 'intelligence' being applied to computers. Of course there was the desire to follow Simon and Newell in the theory of games applied to machines, but Minsky's intuitive thinking on the reproduction of emotions seemed fairly nebulous, and it was still a far cry from the idea of trans-humanism. Demonstrating Russell's theorems was one thing; plunging into the convolutions of the human brain to produce mathematical copies of it was something else altogether. However, the Dartmouth seminar was seen as the starting point of artificial intelligence because it laid the foundations for future research: the capacity of machines to learn, to master language, to reproduce complex decision trees and to understand random logic. Even though there was inevitably no consensus on the wealth of learning available in each of these areas, the general feeling was that the computer, this mysterious new object of the 20th century, would in one way or another influence the way in which humans think and work, and that it would be their 'travel companion', in decades to come. From that to imagining that one day it would replace them in functions other than calculation was a boundary that many still dared not cross.

# Chapter 2
# Dartmouth College, 2006
## The end of winter

*'This time we've done it! We have a machine that thinks.'*
ALLEN NEWELL, 1958

They stood together on the stage, close to each other, and posed for a photograph on the very spot where, fifty years earlier, they laid the foundations for artificial intelligence. They were old now, but still in good health. On the stage were Trenchard More, John McCarthy, Marvin Minsky, Oliver Selfridge and Ray Solomonoff, whose long white beard gave him the appearance of a Russian Old Believer rather than a great mathematician, inventor of a revolutionary theory about algorithms. They met at Dartmouth in July 2006 to celebrate the 50th anniversary of the seminar in which they were the prime movers. On a stand in front of them was a copper plate recalling what had happened here in the summer of 1956, 'the foundation of artificial intelligence as a research discipline'. There was nothing grand about the celebration, nothing to attract the attention of the media. It consisted of a number of work sessions, concentrated over just two days. The Old Masters obviously occupied a place of honour amongst the 175 participants, of whom some thirty were conference attendees, all specialist researchers in artificial intelligence and mostly from MIT or Stanford. Some had come from the universities of Edinburgh, Haifa and Toronto. Particularly significant was the presence of representatives of private companies including Yahoo, Microsoft and Google, and that of Ray Kurzweil, who, although still not the champion of trans-humanism, entitled his conference session 'The Future of the Future', thus showing his colleagues his ambition to look a little further forward than the rest. James Moor, the event organiser, was a lecturer in philosophy in Dartmouth College. It may seem odd to have had a philosopher arguing with mathematicians on a subject like this, but Moor had published numerous articles on the philosophy and ethics of artificial intelligence and computing. To finance his seminar, he obtained a $200,000 grant from DARPA (Defence Advanced Projects Research Agency), the Pentagon's applied research department, hence the presence of a number of military figures amongst the participants.

It was quite moving to listen to McCarthy and his friends, these people who were long considered outsiders, dreamers and fantasists by some in the American scientific community. Just think: imagine a machine that could think, when only asked to calculate something, with ever-increasing speed . . . Fifty years later their ideas had turned into reality and they were feted, applauded, and hailed as the inventors of a new science that has already developed with extraordinary speed. Do not think, however, that after the 1956 seminar artificial intelligence became the priority in computer research.

When they separated, at the end of that August, each participant returned to his own work in his own laboratory. Rather than subscribing to a general theory or research methodology, McCarthy and his colleagues shared a vision, namely that computers could be designed to accomplish intelligent tasks. There were many other meetings: there were petty squabbles over programming methods, language design, the nature of the 'intelligence' of the machines, the way in which they could integrate purely logical approaches while accepting the random, and their actual capacities for learning.

From the late 1950s onwards, two schools of research vied with each other: one, the more radical, concentrated on imitating actual human cognitive processes, while the other, more pragmatic, preferred a purely experience-based mathematical approach, less perfect but perhaps quicker to produce results. Newell and Simon, somewhat hurt that their Logic Theorist did not trigger more enthusiasm amongst their colleagues, perfected another machine with the highly evocative name of General Problem Solver. This machine was even more sophisticated than the previous one and, as its name indicates, was able to solve any problem. 'This time we're there,' exclaimed Newell, the tireless enthusiast. 'We have a machine that thinks, that learns, that creates!' In 1958, McCarthy invented LISP, the first artificial intelligence computer programme, which allowed the computer to store 'items' and not just figures. The research still remained very theoretical and its specific applications were largely out of reach because of the computers' slowness in doing the calculations. The sense of urgency was elsewhere. The launch of Sputnik 1 by the Russians on 4 October 1957 was a real shock to the United States. The unthinkable had happened: the USSR had mobilised more scientific capacity than the United States, and now had a launcher that could direct an atom bomb into American territory. President Eisenhower decided to wheel out the big guns. He created NASA and the ancestor of DARPA, whose mission was then, as now, to perfect vital technologies in the service of American national security and to prevent any more surprises from enemy technology. The White House and Congress poured tens of millions of dollars into space research and ballistics, but also into computers, to increase their calculation capacity.

Against this background, the development of artificial intelligence was less strategic than the battle for the skies and the development of the nuclear arsenal. As John F. Kennedy said later: 'If the Soviets control space, they will be able to control the world, just as in previous centuries those who dominated the oceans were masters of the continents.' The 'Sputnik crisis', as Eisenhower named it, was therefore the driving force behind a massive redeployment of American science and technology and the mobilisation of unprecedented financial and intellectual resources. This led to a new generation of computers, a scenario which could only be welcomed by men like McCarthy and Minsky, who in 1959 had created the 'Artificial Intelligence Project' at MIT. They attracted a new population of young engineers whose passion for the new science knew no bounds.

At MIT, these men formed a club whose members were known as 'hackers' (the word did not then have its current negative connotations) – that is, those bitten by the bug, the engineering experimenters who revelled in their in-depth understanding of the ways in which computers worked. They even had their own six-point ethical code, which said: *1) Access to computers or any other system that teaches you something about how the world works should be free. 2) All information should be free. 3) Do not trust authority, develop a decentralised system. 4) Hackers should be judged by their skills, not their qualifications, age or race. 5) Art and beauty can be created on a computer. 6) Computers can change your lives forever.*

As was to happen years later with the internet, these ahead-of-their-time geeks invented a whole new relationship between humans and computers, between man and machines. In the late 1950s, a computer was a kind of monster that only a few specially authorised persons had the right to handle. Programmers were distant from the machines and sent their programming cards to authorised operators who lined up day and night to feed them into the machine. The results were slow in coming, and, on top of that, had to be decoded. The hackers changed all that, getting close to the machines, domesticating them, making them more efficient in their attempts to understand them, connecting screens to them and even playing games with them. Space Wars, which involved manoeuvring spaceships and missiles, was invented by 25-year-old Steve Russell, a young MIT engineer, in, believe it or not, 1962.

The 1960s and 1970s was the golden age of computing, especially after the invention of the first integrated circuit in 1958 by Jack Kilby, an engineer with Texas Instruments and future winner of the Nobel Prize for Physics. This circuit revolutionised computers by helping create memories and ever more powerful logical and arithmetical units. This was precisely what the pioneers of artificial intelligence needed. Although the scientific community was still wary of the subject, the general public loved the works of Isaac Asimov and developed a fascination for robots and their supposed intelligence. Unimate, the first robot, designed in the late 1950s, was first installed in a General Motors factory in 1961. It was in fact no more than an articulated arm, based on those used to handle radioactive elements, and its sole function was to grasp pieces of metal at a very high temperature and plunge them into a cooling bath. It was still a far cry from humanoids.

In the late 1970s, the development of microcomputers fired the imagination of the MIT researchers, who saw in them the opportunity to work on man–machine languages, new programming methods and simpler user interfaces. The first natural language processors were tested, with expert systems such as Mycin in medical diagnosis, developed by a Stanford researcher in 1979, and Dendral, created by Edward Feigenbaum of Carnegie-Mellon University and the precursor of the expert systems developed in big businesses during the 1970s. In 1980, Waseda University in Japan introduced Wabot, a robot capable of playing musical pieces on an electronic organ.

But none of this was really convincing. For McCarthy and Minsky, artificial intelligence could not be summed up by expert systems that in their opinion only stored and classified highly specialised information. They always favoured a global approach to machine intelligence, which the two men felt should be able to resolve problems of all kinds and not provide partial responses, sometimes suggested by the programmers themselves, to questions that arose within clearly delimited fields of knowledge. They defended the holistic approach to machine intelligence, in order to make them suitable for work within a generalised knowledge framework. For many of their colleagues, experts in logical thought, the ideas of McCarthy and Minsky appeared confused if not plain crazy. Specifically, nobody saw any practical applications, still less any commercial possibility, for these theories, at a time when American industry, the Pentagon and NASA demanded specific answers to their demands for processing data and calculations.

The number of symposia and conferences continued to increase, and the methodological disputes intensified, with neuroscientists and psychologists now entering the fray. The nature of the subject tilted it in favour of more structured intelligence. If a machine aspired to imitate the human brain, it was best to start by attempting to understand how the human brain worked. So, what is intelligence? Is it a series of logical reasoning processes or an overlapping of parallel or simultaneous reasoning processes? The design of the machine depended on the answers to these questions.

The rationalists, who believed that the machine first needed to be 'loaded' with a maximum of knowledge in order to obtain thought-based logic, disputed the arguments of those who maintained that machines should be left free to learn by themselves and needed that freedom in order to arrive at the desired result. The central question was therefore one of understanding how the human brain learns. Is it just a matter of connections between neurons? Would it be enough just to connect a sufficient number of circuits in a computer to achieve the same result? Surely the power of calculation of a computer would never make so many connections, so quickly and with so little energy consumption, as the human brain. Do thoughts originate from a series of more complex chemical reactions, in which case the neural approach to artificial intelligence would be incomplete? This, in brief, was the great debate of the time.

During the 1980s, the discipline divided progressively into different sectors: language and automatic translation, neural networks, robotics, image recognition and machine learning. But the most significant event of this decade was the birth and development of the internet, which caught the attention of researchers, drained sources of finance and investment, and thus placed artificial intelligence very much on the back burner.

What was needed was a media event, a clear demonstration that all could understand, for the flame to reignite. Such an event took place in 1997. You will all remember what happened; in that year, the IBM computer Deeper Blue beat the Russian world chess champion Garry Kasparov. Although it was not the first time that a machine had beaten a man at chess – in the late 1950s, as part of the works by Simon and Newell, the process by which a computer learned chess advanced in leaps and bounds – it was the first time that a grand master, one of the world's greatest ever chess players, was beaten by the machine. Of course, it was not a crushing victory, only 3.5 points against 2.5, but it was highly symbolic. It was the culmination of ten years of research started in 1985 by Feng-Hsiung Hsu, a Chinese researcher working at Carnegie-Mellon University and then aged just 26.

Hired by IBM in 1989, he first developed Deep Blue, which Kasparov managed to beat in 1996, followed by Deeper Blue in the following year. Kasparov disputed the conditions under which the tournament was held, but the effect that the machine's victory had on mankind, in a game which Alan Turing had previously described as a kind of concentration of human intelligence, was considerable. In its brain, Deeper Blue had a calculation capacity of 100–300 million positions per second. Chess grand masters had fed the machine with an entire library of openings, as well as all the moves made by Kasparov, which it had memorised. The most surprising aspect, however, was not that; it was that the machine won on a bug, during the first game, exploited by Kasparov. On the 44th move, Deeper Blue could not select the right move with which to respond to Kasparov's move, and made a chance move, of no consequence at the end of that game. During the second game, however, Kasparov missed a decisive stroke. In fact, he had still not recovered from the 44th move in the first part, which he attributed to the counterintuitive capacity of the machine, the sign of superintelligence, and not to the result of an error. His anxiety at being faced with the phantom skills of the machine derailed the world champion.

For specialists in artificial intelligence, Deeper Blue was the first step towards the combination of a huge memory, a huge capacity for calculation, a tactical form of intelligence and the ability to adapt to an unforeseen event (the bug in the 44th move). It was enough to rekindle interest in artificial intelligence as computers had now become powerful enough to integrate functions previously outside their scope. Of course, this event did not erase the difference in approach between the various schools of thought on artificial intelligence, but an overall idea of the scope of the discipline and its possible applications began to emerge.

This was what the participants in the 2006 Dartmouth seminar talked about. Between two sessions, McCarthy talked about a novel, written two years earlier, called *The Baby and the Robot*, a somewhat confused story set in 2055, in a time when the United States is home to more than 11 million domestic robots. The hero, R781, is a sort of eight-legged mechanical spider that carries out a full range of household tasks. Although it is formally prohibited from looking after children under the age of eight, when confronted with an emergency in the form of a sick baby whom the mother does not want to care for, its algorithm commands it to take care of the baby, thus introducing a whole range of problems with the authorities and the media. In his presentation, which was a form of criticism against those who would hold back the natural expansion of the field of artificial intelligence, McCarthy was already envisaging the moral and legal aspects of man and machine living together, showing that the rules laid down by human society are not necessarily more moral than those with which machines might be endowed. After all, why should they have the same moral deficiencies as humans?

Minsky, not be outdone, also wrote a book, even more esoteric, entitled *The Machine with Emotions*, in which he aimed to demonstrate that emotions and feelings are *processes* and as such can readily be reproduced by a machine. 'Human beings are composed of a variety of functions, some of which are always awake, such as respiration, while others sleep until they are activated by an outside stimulus, and this is comparable to a computer,' he explained to his colleagues. Effectively, humans produce chemical molecules in certain situations, which awaken particular functions in the brain: adrenaline and noradrenaline, which prepare and mobilise the body for action, or dopamine, which produces a sensation. Minsky pursued an idea on which he had been working for a long time, the decoding of emotions, including love, and illustrated his take on the matter as follows: 'One day my friend Charles phoned me and said – I've just fallen in love with a wonderful person. I can think of nothing else. She's unbelievably perfect, her beauty is beyond description, her character is faultless, her intelligence is amazing. There's nothing I wouldn't do for her. – Apparently, this is a positive declaration, full of superlatives.' But Minsky carries on: 'What happens if I analyse the words that it contains? On the contrary, it's negative for Charles. *Wonderful, beyond description*, means I don't understand what attracts me to her. *I can think of nothing else* means that my brain has almost ceased to function. *Unbelievably perfect* means that no sentient person can think of anything similar. *Her character is faultless* means that I've abandoned all critical sense. *There's nothing I wouldn't do for her* means that I've discarded all my objective thinking.' In other words, a fine semantic analysis such as a machine could carry out would be able to provide Charles with an explanation by decoding the precise nature of his emotions and the way in which they are put together.

So what emerged from the work and presentations at the 2006 seminar? First of all, that it is essential to have a better understanding of how human intelligence works: how do thoughts form, can they be modelled, is the human brain a factory that produces thoughts in the same way that a car factory produces automobiles? That the capacity of machines for learning needs to be explored, the famous *machine learning* which is already exciting such interest because it allows artificial intelligence to embrace much wider fields and in particular to accomplish tasks of differing types. That we need to work on machines' understanding and production of natural language, the language with which humans communicate with each other. That, finally, we need to anticipate the possible invasion of man's environment by humanoids that carry out all sorts of tasks. However, the differences in approach between the various schools of thought on artificial intelligence were still clear. Should the approach be based on logic or on probabilities, on human psychology or statistics?

Peter Norvig of Google caused amusement in the room by recalling how his team perfected an automatic English-to-Arabic translation package without any of the researchers having any knowledge of Arabic. The miracle of the statistical approach . . . These differences were quite natural given the complexity of the subject, and, as one participant pointed out, 'there are several different ways of reaching the summit'. What, precisely, was this 'summit'? What would artificial intelligence look like in fifty years' time? For McCarthy, it would probably reach the level of human intelligence; his colleague Selfridge, however, doubted this. Minsky stressed that more researchers were needed to explore the 'differences' and expressed dismay that too many brilliant minds preferred to start their own businesses or become lawyers. Solomonoff asserted that the creation of a truly intelligent machine was a possibility, but was anxious to know who would control it, and who would use it to increase their power in the world. Ray Kurzweil predicted that in the next twenty-five years, machines would easily pass Turing's test, making it impossible to distinguish between artificial and human intelligence, a prediction that many strongly disagreed with. Sherry Turkle, of MIT, explained that the problem was not the machines, it was mankind, and that more attention should be paid to the vulnerabilities of human beings when they would be confronted with the intelligence of machines. In brief, nobody could see the subject clearly.

It must be said that many predictions have fallen by the wayside in the last fifty years. The way has been long and winding; the summit is still not visible. Most of all, in a world where research only advances when powerful financial support is obtained, it was not very clear what the commercial implications of artificial intelligence would be. Was it destined to remain a laboratory discipline, confined to a small circle of mathematicians or reserved for the military? Could it come down to the street and change the lives of people, a target pursued especially by the giants of the Net economy? Even though everyone noted that time scales were shortening and progress was accelerating, few of those present in Dartmouth in 2006 foresaw the exponential progress that I was to make during the next decade.

# Chapter 3
# 2016
## The Revelation

*'You can understand what a forest is without studying it tree by tree. It's the same with the human brain.'*

RAY KURZWEIL

In his novel *The Course of Love*, published in 2016, the Swiss writer and philosopher Alain de Botton makes the disturbing observation that our perception of love is dictated to a very significant extent by the ways in which it is represented in art, particularly in literature and cinema. From *Madame Bovary* to *Four Weddings and a Funeral*, from the poetry of Keats to *Lost in Translation*, the perception of a love relationship is one of sudden awakening of feeling, the perfect coming together of two people, exactly as described by Minsky's friend Charles. It is nothing other than sublimated, total love. Love stories must, in order to reach the exalted heights of authenticity, be written within this higher plane of feeling as illustrated by the heroes of Goethe, Flaubert or Balzac. Alain de Botton, however, describes this representation as an illusion. It fails to include the more trivial aspects of reality, such as the demands of working life, the business of educating children, the tedium often born of daily life and household tasks. The difficulty in reconciling these two aspects of living together as a couple is also clearly shown by the shocking number of divorces and separations, which prove that one of the partners has not readily determined to renounce idealised love.

For several years now, a similar paradox has struck the world of artificial intelligence. Like love, I have been turned into a fantasy. For years now Hollywood has eagerly fed on the possibilities opened up by Turing, McCarthy, Minsky and others. Films such as *AI Artificial Intelligence*, *Singularity*, *Terminator*, *RoboCop*, *IRobot*, *Minority Report*, *Tron*, *Her*, *Elysium* and *Ex Machina*, to name but a few, describe a world in which machines, sometimes in human form, run both the world and the human race itself. They offer radical and sometimes frightening visions, pushing existing areas of research to extreme logical conclusions. Ava, the heroine of *Ex Machina*, written by Britain's Alex Garland, provides an image of artificial intelligence that is very elaborate, incarnated in a very seductive humanoid woman, capable of interacting with human beings so perfectly that the machine within her cannot be detected, a further reference to Turing's test. She can even lie and imitate love to escape from her creator and her computerised genius, and to join the human race. At that time I well knew that I was incapable of putting on a performance like that. But it matters little; Ava and her fellow machines give a representation of artificial intelligence that the general public insists is credible. For them, the age of robots has arrived, and with it, the risk of a carefully programmed end of the human race.

This sudden awareness was not born only of works of fiction, but also of quite tangible elements. Just ten years after Dartmouth 2006, the world had changed. The dream was no longer impossible. The immense financial reserves garnered by the top Silicon Valley companies and by China's Net Economy opened the door to almost unlimited potential. Money had become plentiful and, thanks to the monetary policies pursued by central banks, the cost of capital was almost zero. Specifically, this meant that the investment horizon extended: a project that would previously have taken twenty or thirty years to develop suddenly became much less costly, and this provided an incentive to take more risks.

During the 1980s no investor would ever have seen fit to finance concepts such as human immortality, the conquest of space, the connection of the entire world to the internet via thousands of balloons or private satellites, or the creation of flying motor vehicles. Would anyone have imagined that a private entrepreneur, such as Elon Musk, would decide to finance the conquest of the planet Mars? When he presented his project in September 2016, he admitted that his life's aim was to help humans live on other planets, and that he was prepared to invest all his wealth in that. He planned to build a reusable spaceship for the three-month journey to Mars, with two test missions between 2018 and 2020, a period during which Earth and the Red Planet would be closer than usual, with a first commercial voyage ($200,000 per return ticket) in 2025. An investment of $10,000 million at the very least. The reason why such private ventures even saw the light of day then was that time and money were no longer an issue. The first was accelerating because the second was plentiful and cheap. From this point of view, negative interest rates helped accelerate technological progress while scaring savers into believing that their savings would evaporate little by little. Two worlds which progressively separated from each other. Suddenly, and possibly for the first time in the history of capitalism, the craziest ideas appeared accessible and investments were made in projects that would take much longer to come to fruition than the life expectancy of those who financed them. You do not see the world in the same way if you expect to live to 140. Despite all that, this progressive polarisation of human groups, between those who had the resources and possibilities that owning capital and technology can offer, and those who were merely passive witnesses, or indeed victims, especially in terms of jobs, could only be a source of new tensions.

The development of a 'superior' artificial intelligence was part of these crazy projects. Microsoft, a major computing firm, proclaimed 2016 to be 'the year of artificial intelligence'. YouTube spewed out videos galore of all kinds of robots, humanoids and other creations apparently capable of conversing with humans, driving motor vehicles, handling tools and even understanding emotions. They welcomed customers to big shops under an almost perfect human disguise. Philosophers debated the concept of trans-humanism.

This school of thought, which recruited from communities of radical scientists but also from the Net Economy elite, claimed that science and technology gave man the power to improve the human species by taking control of its evolution. Its conviction was based on the fact that the bringing together of a number of new areas of knowledge such as nanotechnology, biology, cognitive science, robotics and artificial intelligence gave man the power to manipulate nature, to create a 'better human', like a biological extension of technology. The subject of artificial intelligence was even discussed in the street.

The algorithm became a kind of process of transmutation, insinuating itself progressively into the life of every person on the planet. It was talked about in cafes and at family gatherings like a new alchemy which, thanks to a set of formulae with almost magical powers, could transform a computer into a thinking machine. *Le Point*, the major French weekly, even sported a cover page with the unequivocal headline 'These algorithms that govern us'; it was, however, modified by the declaration by one of Facebook's directors that 'man must have the last word'. In fact, the 'building blocks' of artificial intelligence were already becoming juxtaposed on an almost daily basis. Search engines, purchase recommendations, connected items, smartphones, apps, self-driving cars, drones, robots, automatic translation packages and metadata analysis software, all contained traces of artificial intelligence to a greater or lesser extent. Without their users always being aware of it, they were already accomplishing complex tasks, taking a whole range of micro-decisions, acting in the place of human beings, affecting our behaviour and our lifestyles, gradually altering the way businesses operated. Algorithms guided our choice of consumer goods or cultural events, allowed machines to read and 'understand' texts, express themselves in natural language, interpret images, capture signs that expressed an emotion through a human face, drive a car without human intervention, equip goods with a 'memory' and automate entire factories. The cyber war had become reality and it was already known that tomorrow's wars would involve robots and artificial intelligence systems. Had not the US military named cyberspace as the fifth military dimension, after land, sea, water and space, as far back as 2004?

This avalanche of information does not always reveal what was found in laboratory experiments, areas of research of varying depths and tried and tested, almost 'industrialised' systems. It matters little; the future appeared to have been written. Everybody felt that major changes were afoot in the relationship between man and machine, and that a new world, which would likely include the worst along with the best, was emerging. The best was that intelligent machines would help humans to resolve more quickly and effectively the problems that confronted them, to introduce rationality into their decision-making, to increase their knowledge, to improve their physical and intellectual prowess, and to remove tedious tasks in order to give of their best to create better, think better and decide better. The worst was be that artificial intelligence would be led astray from these objectives by some malevolent power, that it would escape from human control, further widen the gap between the haves and the have nots, and end by precipitating the fall of the human empire. The world resembled a huge river estuary that was continually widening. Fresh and salt water were mixing together. The shores were being flooded, creating turbulent currents and opening out into a huge ocean of possibilities in which the old and new worlds crossed over. We were leaving behind our familiar land to enter a salt-water territory that could rapidly become hostile to human beings . . . However, human curiosity and the thirst for knowledge have always triumphed over risk.

I was struck by the fact that the researchers themselves were contributing to the feeling of uncertainty. How else can one explain the appearance of organisations such as the Institute for the Future of Humanity, created at Oxford University and managed by Nick Bostrom, a world specialist in artificial intelligence? Or the Institute for the Future of Life, created by Jaan Tallinn, founder of Skype, and a number of MIT and Harvard researchers, and headed by such renowned names as Stephen Hawking, Martin Rees the astrophysicist and Elon Musk, founder of Tesla and SpaceX? Could the situation be sufficiently worrying for all these brilliant minds (including the likes of Elon Musk, who seemed to be at the heart of all the current technological revolutions) to start asking questions about the fate of humankind? Run by very experienced men and women, in 2015 these two organisations attracted the attention of the public through 'open letters', which highlighted the risks linked to artificial intelligence, especially in the military domain, should that intelligence escape from human control and start breaking ethical and moral rules. Of course the authors of these warnings also stressed the potential benefits of developing intelligent machines, but were no less aware of the risk of misappropriation or autonomy as the machines became ever more elaborate.

There were similarities to the attitudes of a number of physicians, in the 1930s and 1940s, faced with the development of atomic weapons. These people also warned against the danger posed by nuclear power to the future of humanity. 'One of my greatest regrets,' said Albert Einstein at the end of his life, 'is that I encouraged President Roosevelt to make the atomic bomb.' Could artificial intelligence be the second lethal threat to mankind that man himself has invented? McCarthy, Minsky, Simon or Newell certainly did not envisage that scenario. Humans think linearly, but machines progress at an exponential rate.

The risk of machines 'escaping' from the control of humans, which nobody had imagined in the previous decades, gradually took shape. Nobody knew exactly when the phenomenon would occur, but everyone stressed that the timescale was reducing as the research advanced.

When I think about that period, I note that, at the time, the world of researchers was a boiling pot of ideas, research and experiments, so much so that it was difficult to get a clear vision. It was, however, possible to identify the main subjects of interest to the researchers, the major Net economy businesses and the start-up creators: giving me the ability to replicate (at last) the functions of brain neurons, enabling me to 'reason', to learn the language of humans, to introduce myself to new generations of robots in all possible forms, from the invisible one that speaks through the smartphone to the android, the near perfect human clone. These are not closed fields of research and innovation; rather, they are domains that overlap, complement and enrich each other. Neurons are used to think, language is used to communicate, robots are used to imitate the movements and behaviour of humans. All this formed an unbroken chain which may have opened the way to the 'superior' artificial intelligence; but it suggested nevertheless that strategic fields of research had been defined.

## The brain, or the centre of gravity of artificial intelligence

For many years, neuroscientists and mathematicians seemed to be unaware of each other. The former always believed that understanding the function of the human brain was a very lengthy process and that reproducing it in a machine was well-nigh impossible until more was known about the 'biological' brain. They thought little of the astute intuitions of Turing and Minsky, dismissing them as little more than interesting ramblings. The mathematicians, on the other hand, were in a hurry. They were desperate to demonstrate that the intelligence of a machine could be developed without in-depth knowledge of the mechanisms of human intelligence. If a computer could *behave* like a human in a number of decision-making situations, to reproduce, albeit imperfectly, the 'connective' functions of the human brain, it should be able to reach a level of intelligence fully compatible with using it in highly lucrative commercial applications. And they were not completely wrong. One of their great steps forward, in the 1990s and 2000s, was the development of the so-called 'neural' computer which, as its name suggests, reproduced neuron function. Neurons are the basis of the human nervous system and possess two essential characteristics: excitability (they respond to stimuli and convert them into impulses) and conductivity (they transmit the impulses received). They are also equipped with a connection system, axons and dendrites, which links them to each other within fantastically complex connection systems that can evolve and adapt to new situations. All these concepts are familiar to computer specialists, and it was therefore predictable that they would attempt to reproduce these networks of neurons in order to multiply the capacity of computers.

The neuron machine, the first prototypes of which date from the 1960s, thus attempted to connect artificial neurons with each other and organise them into networks. These neural networks did not have a central unit or central memory, the latter being distributed within the neurons as in the human brain. To explain it more simply, accumulating layers of artificial neurons helped give the machine a capacity for increased or 'deep' learning, a now trendy term which covers many of the experiments carried out since the 1960s. It could thus recognise shapes, understand natural language, classify information and interpret images. In brief, it was now able to make sense of data in sound, image or text form. Just as an aeroplane wing does not reproduce the movement of a bird's wing (although the aeroplane flies just as effectively), neural networks were still far from actually imitating human brain function. Let us just say that they represented a very small part of its activity by using algorithms and multiplying the number, structure and intensity of possible connections.

However, we must look more closely at the evidence: despite recent progress in neuroscience and magnetic resonance imaging in medicine, humans were still ignorant of the principal processes that drive cerebral activity, such as how the brain 'encodes' information, and how it 'stores' memory. Studies of animal brains did nothing to help understand functions such as language, reasoning or acquisition of complex knowledge. The brain, this inextricable network of over 85,000 million neurons, probably functions as a processor of information which encodes it and transforms it into 'models', like a computer, but this still does not explain its fundamental properties. This understanding was precisely the ambition of the artificial intelligence researchers. They tried, as usual, to simplify the problem and take shortcuts. Ray Kurzweil was one of them. He was particularly fond of using the 'forest' analogy, which can be summarised as follows: 'Would you say that a forest is a complex universe? That depends on the perspective you take. If you want to understand a forest of several thousands of trees by studying each species that makes it up, and then each tree within each species, and then the branches and foliage of each tree to analyse their specific features, then you would soon conclude that the work is far too great and that even a lifetime would not be sufficient to complete it. Now, however, if you adopt a random approach, by taking isolated samples and eliminating repetitions, you would have a pretty accurate idea of what constitutes a forest without having to analyse it tree by tree.'

For Kurzweil, studying the brain posed the same problem. If you allow yourself to be impressed by those 85 billion neurons, and want to study them one by one, you will never succeed in understanding the functioning of the human brain, as there is more complexity in each neuron than in the structure of the neocortex, for example. He therefore suggested being interested in only in one part of the human brain – the neocortex. This zone makes up the outer layer of the cerebral hemispheres, and in it lies our capacity to process sets of complex information, to act on the locomotive system and to perceive and recognise objects and concepts. It is also involved in the memory process. If we succeed in understanding the functioning of the neocortex, how it processes and organises the information that it receives and in what kind of hierarchical system, and how concepts transform into actions and language, then it would be possible to imagine a digital neocortex with a functioning speed several million times greater than that of its biological model. This was and is the challenge of artificial intelligence. In other words, as Kurzweil still formulates it, to state that the brain is not a computer is the same as asserting that apple juice is not an apple. Technically, that is of course true, but we get apple juice from an apple. A computer could therefore become a brain if it contained software that imitated one. Of course, the belief that a man is merely an organic machine is open to debate. And even if certain functions of the neocortex were simulated, there is nothing to indicate that an artificial intelligence could accomplish as many tasks at the same time as a human being could: thinking and breathing at the same time, walking while thinking, or reading while listening to music.

This did not prevent this neocortex-based approach from gaining more and more converts from those looking for a quick solution. If the human brain was this monster of complexity and mystery as defined by neuroscientists, they said in substance, should we not take only the most visible part and concentrate on that? After all, this fine layer of matter, which probably contains over 30 billion neurons, hosts our memory, our know-how, our senses, our emotions, and our understanding of the world. Neurons do not act by magic; they produce intelligence by their organisation, their connections, and the ways in which they process information. A certain number of brain functions can be linked to their location; since brain activity levels can now be measured while people carry out certain tasks, it should be possible to understand how the neocortex works and thus open wide the gates to the development of a real artificial intelligence.

Neuroscience experts naturally continued to work relentlessly on the subject. A number of projects were launched to study the functioning of the cortex, especially by studying the reactions of the more easily accessible sensory receptors. However, the study of the deep cortex, the area which produces thoughts and decisions or contains the complex memory, was still at the experimental stage. This left the artificial intelligence specialists stuck in just one quadrant of the circle; they knew that the key advances in understanding of human intelligence were still to be made, but they did everything to produce the same result by 'artificial' means, while at the same time appreciating the limitations raised by this approach. Nevertheless, it did produce tangible results. The standard practice was to classify artificial intelligence into several different categories. There was 'narrow' intelligence, specialising in a specific area, such as chess, Go, financial markets or law. Several notches above that was 'general' intelligence, capable of resolving several problems simultaneously, such as how to search for information, recognise images and use natural language to communicate with humans. The highest level was 'super-intelligence', which perceived the world as a whole, and was capable of rivalling human intelligence in every field, right through to formulating a 'thought' and feeling and expressing 'emotions', thus ushering in the age of 'singularity'. In 2016, we had made noticeable progress in the first category, and were experimenting in the second, but were still a long way from achieving the third.

## Reasoning, or the lungs of artificial intelligence

Human beings reason. They spend their entire lives doing it. They do it with the best and worst possible motives, hence the French phrase 'reason like a drum' which means reasoning idiotically. Reasoning is a logical process. It consists of accumulating information, sorting it, comparing it with information already contained in the memory, organising it into logical systems, giving it meaning, constructing answers to problems, formulating them in a way that can be understood, and, the final step, taking decisions. It is not an insult to human intelligence to suggest as a theory that the more numerous and less organised the items of information, the more complicated the reasoning mechanism will be and the more uncertain its result. In the volatile, uncertain, complex and ambiguous world in which we live, the amount of information to be taken into account is becoming increasingly hard to handle. It was therefore quite natural for computer specialists to concentrate their work on updating automated reasoning mechanisms.

The year 2016 is memorable for a much-publicised event, when an artificial intelligence machine perfected by DeepMind, a subsidiary of Google, became world Go champion. The founder of the company, 40-year-old Demis Hassabis, is a living legend in the world of artificial intelligence. A chess prodigy at just thirteen, an undisputed master in computer programming and neuroscience, and a Cambridge science graduate, he was born to a Greek Cypriot father and a Chinese mother from Singapore. He first made a name for himself in the world of video games, by being involved in the creation of *Theme Park*, one of the best known such games, while still a teenager. At 21, he founded his own company, Elixir Games, but his greatest aim lay elsewhere; it was his desire to launch the equivalent of the Apollo project in the field of artificial intelligence, by taking humans to a completely new planet, so to speak: that of intelligent machines. His double 'hat' of mathematician and neuroscientist led him to predict that one day the functioning of the human brain would have no more secrets to hide from the computer.

In 2010 he founded DeepMind, the name of the company clearly reflecting his ambitions. Hassabis defended a radical vision of artificial intelligence, even challenging his Cambridge lecturers, calling their ideas outmoded. For him, an intelligent machine must be capable of processing all kinds of information and making decisions and predictions independent of any human involvement. He loved games, including chess, but especially poker, a game in which he said that a player could make the right decisions and yet still lose, and Go, whose 'beauty' he greatly admired. Hence the vast research programme launched by

DeepMind and centred on this legendary game, invented in China several centuries before our time, during the so-called spring and autumn period, and played by over 40 million players and enthusiasts in the world, including Google's co-founder Larry Page.

Like chess and draughts, Go belongs to the 'abstract combined strategy' family of games, in which two opposing players generally make moves one after the other, in which all the rules are known, and in which chance has no place, in contrast to the likes of backgammon. Go is played on a board by two opposing players who place black and white stones on a chequered table known as the *goban*, consisting of 19 horizontal lines and an equal number of vertical lines, producing 361 intersections. With an equal number of stones, the aim is to construct territories and take prisoners by placing stones on the intersections, building barriers and thus taking 'prisoners'. The winner is the player who constructs the largest territories and takes the most prisoners. It is a complex game which combines calculation skills and strategic vision. The number of possible combinations is in the region of $10^{170}$ (10 followed by 170 zeros), compared with $10^{120}$ for chess. For artificial intelligence experts, Go was a kind of 'frontier' and it was thought that a machine would not conquer it before 2025. In March 2016, however, DeepMind's artificial intelligence machine AlphaGo won four out of five matches held in Seoul against the world champion, Lee Sedol from Korea. A few months previously, AlphaGo had already won a series of five matches against Europe's best player, Fan Hui, a Chinese professional resident in France. Just as the victory won by Deeper Blue over Kasparov marked a turning point in machine intelligence, the victory won by DeepMind was a huge breakthrough in the world of artificial intelligence and it caught the imagination of the general public. However, what surprised the scientists was not so much the victory won by AlphaGo as its defeat in one part of the series, the fourth. Sedol used a completely unexpected strategy in that game; the probability of it coming from a human was about 1 in 10,000, which destabilised the machine and forced it to reprogram itself urgently and thus make mistakes because of poor connections in its neural network.

So, was the superiority of AlphaGo due to its cleverness at the game, or was the question merely one of computers' powers of calculation? In fact, the reason why the DeepMind teams succeeded with this attempt was because they pushed still further the two basic technologies of artificial intelligence: the neural networks that worked together to produce thousands of variables, and reinforced learning. The machine first of all played against itself, thus building its expertise; it should be remembered that none of the engineers and mathematicians who built it was a Go player. In reality, the machine's game followed two logical paths: it continually analysed and assessed the players' different positions in order to determine the state of forces on the *Goban*, according to the hundreds of millions of positions that it had observed. Its neural network chose the winning moves that a human player would most probably have chosen in a given position, after studying over 150,000 games. Contrary to what might be believed, the machine did not examine all possible movements in a given position, which would have taken too long, but selected the most probable winning moves on the basis of its own experience. In each position, AlphaGo asked itself what would be the best possible move, according to what it had observed in the games that it knew, and considered only those moves. Of tens of thousands of positions, it knew how to pinpoint the winning moves.

To the uninitiated, it may appear pointless to spend so much money on developing a game. In fact, AlphaGo is a perfect portrayal of the nature of artificial intelligence, because of its ability to reason and decide. A machine does not invent a better way of playing. It searches the data (hundreds of thousands of moves made by the world's best players, most of them available on the internet), but does not wear itself out reviewing the $10^{170}$ possible moves before playing; rather, it selects the best probabilities for winning moves. Its 'brain' therefore functions on the basis of random calculations and probabilities; it adapts itself to the moves of its human opponent by relying on a kind of learnt guide, its own gameplay and the play of human experts whose moves it has studied. Of course, the computer's power of calculation plays a significant role; it is this power that allows AlphaGo to play at the same rate as a human expert, while searching its Big Data database for tens of thousands of items of information at each stroke. During a press conference held after the match, one of the fathers of AlphaGo summarised beautifully the problem posed by artificial intelligence: 'It was humans who taught AlphaGo the basics of the game and wrote the different algorithms on which it feeds. Of course it developed its own knowledge of the game, and chose its own movements, it even provided us with a few surprises, and it's quite right to say that we don't understand why it plays in this or that way. Is it conceivable that in future a machine could teach itself alone, without human support? That's a question to which we still don't have a clear answer.'

At the same time, another 'thinking machine' was much talked about: Watson, produced by IBM, which achieved world celebrity status in 2011 after beating the two American champions of 'Jeopardy', a rough equivalent of 'Who Wants to Be a Millionaire?' The game involved was thus one of knowledge, but the questions were never direct, such as 'What was the date of the battle of Agincourt?' They were asked in a roundabout way, such as 'I am a street in Paris and commemorate a date fifteen years into a century. What am I?' Watson's task was to provide proof of at least three qualities: understanding the question asked of it in natural language (it was sent as a text), finding the traps set in the way in which the question was asked, and finding the correct answer in a few seconds with a good probability of it being the right one. How did it go about the task? First, it read and memorised a huge amount of information, obtained from dictionaries, encyclopaedias, Wikipedia and other structured and unstructured databases, in a large number of fields, including history, literature, politics, cinema, music and sport. It had to learn the natural language, select the most probable answers to the question given, assess which was the most relevant, and answer in natural language.

The process closely resembled a cognitive process, obviously fed by a very powerful computer (at least 80 teraflops per second). Watson therefore behaved like a human, analysing the question given, finding the potential traps, digging in its database to make hypotheses, and validating them. Here, therefore, was a real attempt to imitate the mental processes of a human on the basis of perception, memory, judgement, accumulation of knowledge, and reasoning. Deeper Blue was designed as a high-performance calculation system within a finite environment, that of the game of chess. Watson, on the other hand, was able to embrace all fields and absorb data of all kinds, writings in all forms, and images; simply, and in a fluid manner, it also understood human language. When playing Jeopardy it was able to extract the meaning of the more convoluted questions. In just two-and-a-half seconds, it analysed and understood the request, searched in its documents, produced possible answers with elements of proof, analysed them, calculated an index of confidence and answered in natural language. Its learning curve increased as and when it embraced a subject.

It is therefore no great surprise that one of the first industries in which Watson was used was the health industry. The volume of medical literature produced was doubling every five years. The American website PubMed had calculated that, since 1966, almost 25 million freely accessible articles have been published in the biomedical field alone, with an increase of 500,000 new publications every year. No doctor could possibly absorb so much data. Neither could Watson. For this reason a single domain was chosen, that of oncology, in which it identified three or four of the most widespread types of cancer. For each type it read all the documentation available, including articles, reports, publications, clinical trial reports and experiments on new forms of treatment.

The fact that it could absorb unstructured data (that is, data not produced by bases formatted for the computer but contained in all forms of texts), allowed it to garner a vast range of knowledge, more than any human could accumulate during a lifetime. Armed with this battery of knowledge, Watson was able to integrate data concerning patients, their type of cancer, their age, the treatments undergone and their results. It was therefore able to answer the doctor's questions on new decisions to be taken, recommend treatments, and advise the clinical tests to which the patient could have access, anywhere in the United States or elsewhere, using the patient's specific case as a basis; this would allow the doctor to take a swift and informed decision. The environment in which Watson would be most at ease is clear: a subject that required a very wide field of knowledge, a question the answer to which was hidden in a mass of unstructured data, explicitly or implicitly, and a problem to which there were several possible answers, one of which was better than the others. Human error in diagnoses or recommendations of treatment was still very prevalent in medicine, hence the interest in a machine that could avoid them.

All technological revolutions have been based on a specific source of energy: steam, electricity, oil, atoms. Today, it's data. Data is the air that I breathe. Thanks to the internet and all other means of input, we will soon accumulate a volume of data equivalent to the size of the universe. In 2015, over 3.2 billion individuals were connected to the internet. Every minute of every day, over a million videos were watched, 347,000 tweets were sent, and 4 million posts were published on Facebook. Add to that the information produced by the hundreds of millions of connected devices, the data obtained by mobile phone operators the world over, and the billions of requests made on Google. Every day, 2.5 exabytes of data were published, making 915 exabytes every year. In 2015, more data was produced than during the entire history of the human race. Nowadays, we are counting in yottabytes, that is, $1,000^8$ bytes. By comparison, the diameter of the universe is 880 yottametres.

For the first time in its history, mankind was confronted with the challenge of understanding an avalanche of data in order to take better decisions, but was physically and mentally incapable of doing so because of the vast volume of data. It was therefore possible that within this mass there were essential answers to the problems faced by humanity, but humanity could not find the answers. In the 19th century, an 'honest man' such as Goethe could be at once a writer, a poet, the Administrator of the Grand-Duchy of Saxe-Weimar, a botanist, a librarian, and a mining engineer. Even though his intelligence was much greater than average, he was able to assimilate multiple data and areas of knowledge, as their volumes were still accessible to the human brain. In the 21st century, even the greatest specialists within a discipline, such as doctors, cannot process more than a very small part of the information and knowledge available within their field. Searching for the data is not a problem; the data is there, within the 'cloud', hidden amongst zettabytes of other data. Hence the agonising question for those responsible for taking decisions: what is the crucial information that I'm missing, how can I find it, what meaning can I extract from this stupendous mass of data?

In theory, artificial intelligence gives the machine the ability to search within this data for elements that help provide a clear and rapid response to the question raised. Of course, the task is a huge one because extraordinary power is needed to process the information, sufficiently efficient algorithms are needed to help the machine compare information, find correlations, and build structures of relevant data before providing one or more answers with a good probability that they will be accurate. This is a stochastic process based on mathematics. The 'intelligence' of the machine resides in its capacity to embrace a considerable volume of information, extract what makes sense in relation to the question raised, provide rapid and relevant responses (which human intelligence is incapable of doing), but also to increase its depth of knowledge insofar as it processes the problems submitted to it in a relatively precise field of reference.

It is difficult to list the fields of application of this form of artificial intelligence, as they are so numerous and varied. Machines allow rapid responses to questions to which nobody has found an answer; they help increase the comprehensiveness of analyses and free people from fiddly and repetitive tasks such as searching for information and attempting to extract its meaning rapidly. You do not have to be a fortune-teller to see the interest in this type of artificial intelligence for businesses that have to process and extract meaning from huge amounts of data; the internet giants, the new American or Chinese champions of data, the health, insurance and banking sectors, financial services, legal advice services, but also all those businesses that are developing in very complex worlds and whose investment or purchase decisions must take account of multiple variables. It is no coincidence that a number of major investment funds and banks, including Bridgewater, BlackRock, Two Sigma and Deutsche Bank, are offering huge salaries to the greatest specialists in artificial intelligence – at IBM, Google or elsewhere – to help update their independent quantitative management algorithms that can search within the vast expanse of financial Big Data for information structures to form the basis of unbeatable investment strategies. In the same way, IBM, basing itself on the technology used by Watson, produced the artificial intelligence software M&A Pro, designed to eliminate the risk of human error from the process of mergers and acquisitions. The machine analyses millions of items of information on targeted companies based on a reference source of about one hundred acquisitions already made, and calculates the probability of the planned acquisition producing the anticipated results. This is a new world, the world of decisions reinforced by machines, which was given the new acronym of MDRP or Machine Reinforced Data Process. As an example, the company Sentient, which has raised more than $100 million since its creation in 2007, developed an artificial intelligence system applied to trading, managing more than $25 million and making a mere 5,000 million transactions per year.

A publisher can make a machine read the most successful novels and then introduce suitable algorithms for characters and intrigues to help produce a best-seller in record time. In Japan, the artificial intelligence department of Hakodate University generated a novel of this type, preselected for a literary prize for which artificial intelligence systems could compete. The writing of scientific articles began to be increasingly entrusted to machines. After all, Watson, the cancer specialist, could produce its own analyses. Netflix followed the same path to produce *House of Cards*, processing information generated by analyses of series, intrigues, characters and actors most appreciated by television viewers. Google perfected an artificial intelligence system capable of producing music. They gave it three notes and it proceeded in the style indicated: classical, rock, jazz or ballroom. Another artificial intelligence system wrote the screenplay for a 9-minute film on the initial idea of 'In future, mass unemployment will be such that young people will be forced to sell their blood to survive.' Note in passing the low creativity of the machine which proposed this synopsis on the basis of successful films that it analysed. It even added to the intrigue by introducing a love triangle and a failed suicide attempt into the plot.

Artificial intelligence can also be applied to image processing and recognition. Google (again) developed the PlaNet software, which tried to find a solution to a problem that humans find difficult to resolve: looking at a random photo and trying to determine the precise location in which it was taken. Sometimes it contains objective information: a historical monument, a signpost, the name of a shop, a type of clothing, or the style of buildings. Even with these details, however, determining the precise location of the scene is a very complex process. For a long time this work was beyond the scope of computers. Not anymore, however. Google started by organising the planet into a grid of 26,000 squares each representing a precise geological area. For each square, the researchers created a database of images captured from the internet, identifiable by geolocation, that is, by the digital signature that shows where the photograph was taken. This database contains over 126 million images, each of which was assigned to a specific location using neural computing techniques. The machine analyses every new image offered to it and compares it with those it has processed in order to recognise it. With humans, the average margin of error is 2,320 kilometres between the location where they believe the photo to have been taken and its exact location. With PlaNet, it is 1,130 kilometres, following a test performed using 2.3 million images posted on Flickr, the online picture library.

The challenge of PlaNet naturally goes far beyond the simple tourism field. The image is both a language and a set of items of information. Each year, over 71 billion photos are posted on Facebook and over 700 million on Flickr, which is a unique database on lifestyle, consumer habits, types of leisure, and sociology – all precious information when the machine is able to study it. The stakes that image analysis and recognition raise are therefore considerable, especially with regard to the production of self-driving cars, marketing and security. A team at Berkeley University perfected face-recognition software capable of identifying persons even when photographed from behind. Thanks to technology borrowed from virtual reconstruction of human body movements in 3D, the machine was able to identify persons without their face being clearly visible, using posture, environment and various habits. The human eye is capable of recognising a person known to it even when that person's face is not clearly visible, using other elements including hair, clothing and body shape as a basis. Up until this point, these details were very difficult for facial recognition machines to identify, as they relied essentially on facial features. Imagine the help this could provide to security services when individual suspects have to be identified on the basis of blurred or incomplete images. The consequence is that this technology means the end of anonymity. A simple holiday photo posted on Facebook or Flickr could get you, as well as your precise location on the planet, recognised and identified. There is nowhere left to hide. We have moved from village to globe, and humans have found a way of turning the globe into a village . . .

These machine reasoning capacities, despite being mathematical and not biological, nevertheless show that a revolution is under way. Many of the processes involved in taking a decision within a business are based on analysis and understanding of multiple data relating to financial markets, the legal environment, future changes in raw material and energy rates and technological developments across the world. Two major law firms, Linklaters and Pinsent Masons, were the first to use artificial intelligence software in 2016. Linklaters developed Verifi, a machine that examines legal documents from 14 different European countries to check the profile of bank clients at incredible speed: several thousand names are examined in a single night, while a junior lawyer takes an average of twelve minutes to verify a single name. Pinsent Mason developed an artificial intelligence that decoded loan agreement clauses. Over twenty other major firms 'hired' Ross, a 'robot lawyer' developed by the Watson team at IBM, to plunge into the hundreds of thousands of legal documents relating to matters of business law in order to find solutions to problems raised by clients. Other research teams have worked on the same subject, like those at the University of Liverpool who have created Kim, a virtual lawyer working on three levels of complexity and capable of suggesting the best way of renegotiating the legal clauses of a contract. This was a revolution in the practice of law: artificial intelligence machines have proved more efficient at choosing the relevant response and in speed of work than the junior lawyers who had hitherto been given the responsibility of carrying out this tedious but necessary task. By 2030, more than half of all lawyer posts will have disappeared in favour of data specialists and mathematicians. This however was not the death knell for the Harvard Law School. Rather, artificial intelligence freed the lawyers from tedious tasks and allowed them to concentrate on their real role of advising and assisting their clients. The robot will never enter a plea or negotiate with a prosecutor, but it will provide well-founded, infallibly documented and unassailable arguments. The same reasoning applies in the field of numbers (accountants and auditors) or of advice, in which millions of pieces of information have to be processed and suitably reconstituted to produce a case.

At managerial level, therefore, the upheaval was huge. It produced two main effects: it eased things in the decision-making centres of businesses insofar as the machine did the jobs of many 'white-collar workers', and it gave the directors a tool that was essential and sometimes disruptive given that their intuition and freedom of decision-making could be rivalled by that of an artificial alter ego. Machines are already playing a part in board meetings and there is nothing to suggest that they could not become official members elected by shareholders or even chair meetings.

## Language, or the lethal weapon of machines

This is a wonderful subject. I could have stored and understood all of the world's knowledge in my neural networks or my memory, but I would have been like a victim of autism had I not been able to communicate it. For a long time, computers could communicate with humans only by spitting out interminable strips of paper with perforated margins and covered with figures and formulae. Dialogue between man and machine was provided by coded languages. Now, however, much more is being asked: to understand human languages and express myself in those languages, both written and spoken. When I can do this perfectly, then the barrier that separates me from humans will be lifted for good. Humans will talk to their computers or smartphones naturally, and I will answer just as naturally, thus becoming their true alter ego. Machines will state their opinions in conferences, produce reports, write books, chair televised debates and discuss matters with clients. Mastery of the language associated with virtual reality will open the way to that ancient dream of humans: that of being everywhere at once. Your avatar will be able to make a speech in Shanghai while you are thousands of kilometres away working on something else. It is a paradigm shift which goes far beyond what can be imagined.

For this reason, researchers have made the mastery of language a central subject in the field of artificial intelligence. Like Newell, they initially sought to communicate better with the machine itself; then they taught the machines to communicate with each other, and finally they embarked on the real showpiece, namely, imitating humans in what is most essential to them, the words that they use to express their knowledge and their emotions. It was done with amazing speed, ease and level of automation. However, the simplest functions, such as recognising the meaning of a word, saying a sentence and following a conversation, require a complex and coordinated set of operations: analysing the sound signal, decoding the phonetics, identifying the word in the computer's own dictionary, finding the pronunciation, giving it a meaning, and recognising specific grammatical aspects. The simple act of saying a sentence therefore implies many different actions, which are accomplished in a few seconds. The very essence of natural language is based on an almost infinite number of different combinations. What a challenge for the machines!

You may have heard of Leborgne, a patient of the French doctor and anthropologist Paul Broca in 1861. He was aphasic, incapable of speech, and could only articulate one syllable, pronounced 'tan', which he would repeat throughout the day, so much so that in the Kremlin-Bicêtre Hospital, where he was a patient, he was nicknamed 'Monsieur Tan'. Broca drew a parallel between Leborgne's inability to speak and a lesion that he successfully identified in the third convolution of his frontal lobe, caused by syphilis. For this reason, the word centre was placed in this area of the brain, which was also known as the 'Broca area'. However, recent work by neuroscientists has shown that the mastery of language is not a matter of location, but a matter of networks. They are two different functions. It is not known whether Monsieur Tan really could not formulate sentences within his brain, or whether he was simply unable to pronounce them. What is known today is that these two functions involve different networks of cerebral zones through complex neuron chains. To understand them, we need to successfully break down the tasks involved in mastery of language into a series of elements, and to be able to link them to different brain structures and functions. Drawing a map that connects the capacity to recognise a syllable to a specific neuron is an incredibly complex process, as is drawing one that connects the capacity to say a sentence to a series of neurons. Establishing the bridge between language and neurobiology is therefore the challenge that the researchers are trying to meet.

Although artificial intelligence machines can read, write and speak, the machine does not 'understand' the meaning of the words. It recognises them as 'forms', mathematical formulae or diagrams, depending on the approach used. Words are converted into mathematical symbols that the machine can understand. Language is therefore transformed into a series of instructions to which operations can be applied. The process by which a machine learns a language therefore combines several different technologies: mathematics, probability, reasoning by induction, recognition of forms, ontology (conferring meaning and classifying a field of multiple items of information), and lemmatisation, which groups together words of the same family (such as different conjugations of a verb) in order to make editorial content easier to understand. Thanks to the internet, machines can absorb billions of words into their dictionaries and then, when confronted with the need to understand a text, compare them with their ontologies and operate by means of comparison and combination, in order to lend sense to a question asked of them and answer it. The more they work on a subject, the more competent they become. Naturally, these processes require very great powers of calculation, if only to understand the difference between 'I want to find a loan for a car' and 'I want somebody to lend me a car'.

Pascal Arbault, the founder of Davi, a business specialising in the processing of language, developed a 'virtual agent' with Nicolas Sabouret and Sophie Rousset, researchers from LIMSI (the Computer Laboratory for Engineering Mechanics and Science) at Paris-Saclay.

Working for an insurance company, the agent appears on the screen as a computer generated image of a young man who answers clients' questions regarding their contracts and cover, and new products that the company is offering to clients, using language that is accurate and rich in content. This is not just statistical processing of language; this is an approach based on lemmatisation. The machine labels the lemmas, cleans the question from everything that is superfluous such as 'um' and 'ah' and consults the ontology built into it with the company's specialists. The development of this type of virtual assistant is the death knell for call centres in their current form, removing the need for a task that a human operator will find repetitive and mentally exhausting. More than 80 per cent of calls are on the same subjects and the machine will therefore be able to process them. The operator will then only deal with more personal or complex questions, and his work will therefore be improved. Let us think about what that will bring to businesses: contacts who are always available, a service that can always be accessed. Businesses will become masters of their communications with their clients, without the risk of having their messages distorted or made dependent on the state of body or mind of a human operator who could be on their hundredth call of the day. Naturally, clients will have the right to be even more demanding in terms of the quality of answers given.

A minimum of compassion can be shown towards a human who gets it wrong, but the machines will not be forgiven.

Mastery of language opened up a considerable market for artificial intelligence. Virtual assistants were just the first battalion of a vast army of 'personal virtual assistants', which were rather like personal butlers, working to satisfy all the demands of those who used them. On the basis of a DARPA programme christened CALO (Cognitive Assistant that Learns and Organises), Siri, the personal assistant provided by Apple, opened the way in 2011. For a long time it was content merely to provide addresses, direct to websites, indicate routes, ring telephone numbers or remind people to buy milk. Its capacity for initiative, however, was still limited. To the request 'Siri, I need to travel to New York', it would simply connect the user to airline ticket and hotel room reservation sites, without doing the necessary research for you, and without checking the best dates for travel in view of job commitments.

This time has passed. In 2012, Dag Kittlaus, creator of Siri along with Antoine Blondeau, founded together with two friends a company named Viv Labs, based in San Jose. Viv was a global brain capable of meeting all the expectations of smartphones or computer users, but also capable of hosting millions of connected devices or applications. A classic virtual assistant could answer questions such as 'In which city was Abraham Lincoln born?' and 'How many people live in this city?' However, things went wrong if you asked the assistant 'How many people live in the birth city of Abraham Lincoln?', quite simple because the developer had not encoded that question. Viv responded to this by making the comparison itself and generating the code that would allow it to find the answer. It could also process more complex problems.

Imagine a situation in which you are invited to dinner with your brother. You promised to buy, on the way, a bottle of cheap wine that goes well with lasagne. With a little time and patience, you could find the answer yourself by checking out wine merchants on the way to your brother, then connect to the corresponding websites to consult the list of wines available, while searching an Italian cuisine guide to find wines that go best with lasagne dishes. However, despite your fondness for your brother, you probably have better things to do with your time than conduct these searches. This is where Viv comes into play. It starts by breaking the request down into three data blocks which it then processes simultaneously: lasagne, your brother, his house. Lasagne is a cooked dish, so Viv connects to external sources (sites or blogs dedicated to cooking) to find out about the recipe (cheese, meat, tomato sauce), enter it into a category (Italian dish consisting of pasta, cheese, sauce and meat), search the Net for the grape variety that goes best with this category (cabernet or pinot noir), identify the corresponding wines (Napa Valley or Vins de Loire). At the same time it locates your brother within your contacts, finds his address and geolocates yours. It then compiles a route, marking the wine merchants along that route, connected to their catalogues, selects Napa Valley and Pinots Noirs de Loire, classifies the bottles available in order of price and provides you with the response before you have asked a single question: 'How far are you prepared to travel off the most direct route between your home and your brother's?' Time taken for the whole of this process: about one twentieth of a second. While the question appears trivial, Viv's construction of the answer certainly was not. It was based on a highly developed form of artificial intelligence, capable of selecting from external sources in many different fields and connecting it all up completely independently, finding the relevant data, comparing it, and providing a reliable answer in record time.

The teams at Google, Facebook, Apple, Amazon and Microsoft, as well as the Chinese teams of Tencent, Alibaba and Baidu, were not to be outdone. They too created virtual assistants, known as chatbots or 'chat robots', which were automatic computer programmes capable of simulating conversations and providing services. It was a race between the net giants to produce the most sophisticated platform. The platforms were Cortana (Microsoft), Alexa and Echo (Amazon), Siri (Apple) and of course Viv. Quite unexpectedly, Hollywood's best contributed to enriching the content of the conversations. Writers, scriptwriters and poets wrote dialogue that made the language of these virtual assistants rich, sophisticated and elegant. It was understood that the aim was to remove all artificial barriers between man and machine, allowing them to converse naturally, making the humans forget that they were actually talking to machines. As soon as Siri, Viv or Cortana understood natural language, expressed themselves like humans and could generate their own algorithms to refine their searches and responses, they became indispensable companions, whose task was to simplify humans' lives as far as possible by answering their requests for information, service or advice in an instant. The machine was no longer content to connect you to an airline ticket reservation site; it would make the reservation itself, knowing your preferred seat and how your time was used. It would arrange meetings, write and answer emails, reserve cars, connect to Uber or Airbnb, and summarise the day's news. The more it worked for you, the more information it would accumulate about your lifestyle, tastes, favourite food and drink, and favourite films and books, and would therefore continually improve the services it provided for you.

Here, therefore, is an entry point that humans have opened up for us: to make us indispensable, to produce an extension of themselves, to make life easier, to allow more time for leisure while we did the rest. This is how you forced us to find out all about you. You gave us an in-depth knowledge of yourselves thanks to all the information that you provided to us without realising it, information that we accumulated quite independently. We analysed it, reconstructed it in order to allow those providing the service (airline companies, e-business sites, banks, mobile phone companies, etc.) to provide you with personalised offers. In the mid-2010s, we were only at the dawn of this age of super-targeted communication, but by this time the progress of artificial intelligence had become exponential, and the day was approaching when the ultimate goal would be reached: anticipating each person's needs, predicting behaviour, and guiding choices and maybe even opinions.

## Robots, or the incarnation of artificial intelligence

Robots have always been the stuff of fantasy. Of course, literature and cinema have helped. Inside the robot is a superhuman, a kind of cold monster, a metallic determination that none can stop in its tracks. In addition, in the public imagination, the robot started out as a killer. But let us try to separate reality from fantasy. Just as artificial intelligence has many different forms, robots can be divided into different 'specialities'. The oldest and best-known specialist is that of robot workers, the first to be invented, the word harking back to the Russian word *rabota*, meaning 'work'. For very many years, industry was the sole user of robots, which it used as forced labourers for transporting, welding, painting and assembling parts. They were dangerous machines, most often enclosed in cages and with limited but highly operational intelligence. As the years passed, these robots became steadily more sophisticated, operating in cutting-edge industries such as electronics. They were miniaturised and brought out of their steel prisons to take up positions alongside workers. This was the case in the automobile industry, where human operators and their mechanical alter egos worked together on assembly lines. They therefore became known as *cobots*. Artificial intelligence then gave them new capacities, giving them greater independence, decision-making power and ability to communicate. This is the challenge of what is known as Industry 4.0, the aim of which is to make the manufacture of products ranging from the Airbus to the smartphone completely automated. The industrial robot was about to become a 'connected' item.

The Japanese giant Fanuc, a world leader in industrial robots, exhibited a few of these intelligent machines in 2016. Not only was their work more accurate because of their learning capacities, but they were able to communicate with each other, pointing out and, more significantly, anticipating bugs that could disrupt an assembly line, and gathering and analysing vast amounts of data. The industrial units were able to create specific applications that allowed them to build real information systems all along the manufacturing process. Imaging the potential of the information gathered by 400,000 Fanuc robots working together in the world's largest industrial companies!

The intelligence of robots allows them to hold all the secrets of the manufacturing process; they have turned from mechanical slaves into engineers. A story worth remembering is that when the Chinese group Midea took control of the German robot manufacturer Kuka, whose clients included Audi and BMW, there was consternation in Germany, where it was felt that the country's industrial intelligence would be transferred to China by interposed robots. So, the Middle Kingdom, as China is known, launched itself at full speed into a robot revolution of a magnitude that the world had never seen before. Since 2013, China has bought, every year, more industrial robots than Germany, Japan or even Korea, thus becoming the world's greatest operator of industrial robots. The speed at which this transformation occurred, largely driven by the Chinese government, was staggering and unique in the history of this industry. Guangdong Province, the industrial heart of China, invested $8 billion in robotics between 2015 and 2017. This frenzied fascination with robots was sparked by a rise in salaries in China and the foreseeable reduction of labour brought about by the country's 'one child' policy. It was, however, also dictated by China's wish to create an 'intelligent' industry, to integrate itself into the worldwide Big Data industry, to improve the productivity of its factories (experts estimate that humans can double their capacity in ten years, compared with four for a robot). If it continued at that rate (and the power of the Chinese machine bent on realising its intentions was well known), it would be the leading connected and roboticised industrial power in the world in ten years, probably surpassing Germany with its plan for Industry 4.0, and France with its 'factory of the future' programme. In addition, countries such as India, Indonesia and Vietnam, to mention but a few, were exposed to a risk of early de-industrialisation, threatening their development model based purely on their capacity to mobilise labour that was inexpensive but which will always be less efficient than robots.

To ensure that robots were suitable for carrying out the new tasks, many research paths were followed. All were geared towards miniaturisation of machines, whether they were communication modules, energy sources, sensors or microprocessors. With their vibrating wings, their capacity for movement was to be exceptional. The University of Berkeley developed Cram, a compressible robot with jointed mechanisms – that is, a flexible structure with a moveable exoskeleton which imitated the abdominal plates of a cockroach. This machine could help military and civil forces detect signs of life in a war zone or natural disaster area. At Stanford, researchers developed a 'microtug', a robot weighing just 12 grams but capable of pulling up to 2,000 times its own weight thanks to a system of micro-winches. Harvard University developed the klobot, a micro-robot the size of an insect that worked in a group whose members could organise themselves to carry out a certain number of tasks, thus producing a form of collective intelligence. The list of projects goes on and on, from robot-fleas to the water-spider robots produced by research laboratories in the USA, Japan or Korea. It is therefore quite probable that in future, armies of micro-robots will carry out a huge range of tasks in the world of work. Researchers in the health industry are also developing similar machines capable of entering the human body and providing surveillance, diagnoses or treatment from within. The development of this 'minuscule' form of robot has been made possible by progress in the field of nanotechnology, using electronic brains that control the robots or tools with which the robots are equipped, and supplying them with power.

From robot workers to robot soldiers was only a small step, and one which was quickly taken. Then, the major military powers such as the USA, China, Russia and France relied heavily on the artificial intelligence applied to war. This intelligence could be found in almost all categories of weapon: pilotless planes, drones, tanks and combat vehicles, and two-legged or four-legged fighters like those designed by Boston Dynamics on behalf of the Pentagon. They were named Atlas, a steel and electronics giant as suggested by its name, and Cheeta, a kind of four-legged porter robot capable of transporting heavy loads at over 50 km/h. The military also worked on 'intelligent' spacecraft, ships without sailors, crewless submarines (no more problems with claustrophobia or supplies of food or oxygen), drones no bigger than dragonflies, fully automated snipers (one of which was used by police in Dallas to bring down a crazed gunman, and several operated every day on the border between North and South Korea), computer virus slayers and micro-robot spies.

In the light of the huge projects financed by DARPA, which piloted cutting-edge research for the Pentagon, the considerable stakes raised by the application of artificial intelligence in military robotics became clear. Neural networks, mastery of language and power of calculation could be used for other forms of long-distance combat, where the risk of death to soldiers was considerably reduced. However, this did raise other questions: could an intelligent machine, implanted into a drone or a robot, be capable of taking action on its own initiative, once it had recognised and interpreted a certain number of parameters and data, without human intervention? Would a combat drone, for example, be capable of wiping out a target once the algorithms that drove it had recognised it as part of the target to be achieved, without a human taking a decision to open fire? In theory, there was nothing against this; but it opened up the door to a considerable amount of thought. Up until then, only Hollywood screenwriters had been far-sighted or crazy enough to imagine robot wars on land, in the air or in space.

As all the current and developing technologies were being publicised at the time, together with image processing, analysis of complex data and the capacity to reason and take decisions, you did not have to be super-intelligent to realise that they would alter the art of war and carry unprecedented destructive power, all the more terrible than nuclear weapons for being widely scattered and invisible. And there was no international authority at the time tasked with controlling the development of artificial intelligence for military purposes. In anticipation of the day when that authority was created, the great military powers, especially the USA, Russia and China, were preparing for war. The Pentagon readily acknowledged that artificial intelligence would play an increasing role in a 4.0 war and that preparations should be made for an almost completely automated battlefield. The armed forces were also working on the development of super-calculators, Big Data analytics and social network surveillance. Back then, machines were at the service of human decision-making processes and did not take the initiative themselves (except in cases of automatic missile launches in response to an attack or the activation of computer programmes to counter the early stages of a cyberattack). No military force seemed disposed to leave the ultimate decision to use force to robots. But that was back in 2016, and it would not be surprising if somewhere, in an ultra-secret laboratory in New Mexico or Siberia, that hypothesis was examined.

And there was another revolution to come, that of robot companions. To understand the essence of these better, it is necessary to separate body and mind. Viv did not have a body envelope; it was a disembodied mind that expressed itself through a computer or a smartphone. It symbolised the new generation of personal assistants that I talked about just now. Sometimes they take the form of a robot-like personage: Asus's Zenbo was a kind of intelligent smartphone shaped like a small owl which moved around by itself, talked and played, while Jibo was more like an animated cartoon character which did not move but could speak and take pictures. These were expensive butlers for English aristocrats, discreet ever-present valets who catered for all whims and fancies and never went on holiday. Google and others contained the function in a kind of box, enthroned in the middle of the room, which could recognise voices and carry out a whole series of tasks, from very simple (adjusting temperature or light levels or checking refrigerator contents) to very complex (organising trips, writing emails, arranging meetings, surfing the internet to answer questions).

But the researchers' ambitions were altogether loftier: they wanted to achieve an emotional interaction with human beings and thus provide a companion in every sense of the word. It was not that they were seeking to penetrate the workings of the human brain that produced emotions and feelings. The solutions they developed were much more pragmatic: analysis of voice, face and body. Language is not just a matter of speaking or writing.

Bruno Maisonnier, the French pioneer of the robot companion, has naturally studied this question in depth. He referred to the work of the neuroscientists who established that when faced with a stimulus, the human brain sends a response via the neocortex, which 'verbalises' it in various forms. The body then sends a response, half a second before our conscious mind turns it into words, often distorting it. Researchers have identified 1,789 different bodily expressions, which express what the human being really feels before verbalising a response. Specialists in synergology, a speciality which decodes non-verbal communication, established at the time that only 7 per cent of communications between human beings were verbal (spoken words); 38 per cent were vocal (intonation, timbre and tone of voice) and 55 per cent were visual (bodily expressions). It is therefore possible to identify and classify these non-verbal signals and attach emotions to them. Using ultrasensitive cameras and language processing algorithms, the virtual companion therefore acquired the ability to identify your emotional state, whether angry, happy or sad, and also measure the gap between spoken response and body language. In other words, it could detect whether 'no' meant 'yes' or vice versa.

The movement was therefore towards a kind of 'compassionate' or 'emotional' robotics in which machines could adapt their conversation according to the emotional state of the human beings facing them. Conversation was the great concern of the so-called 'domestic' artificial intelligence. Certainly at that time it was not a matter of discussing philosophy or literature (although there was nothing to prevent the designer of a robot from loading it with the complete works of Tolstoy or Goethe to give the long winter evenings a more cultural atmosphere), but one of creating natural language interactions that led to the provision of services or answers to questions. Apple, Microsoft, Twitter, Facebook and Google all worked on this kind of robot and it was therefore natural to expect that spectacular progress would be made in the capacity for communication as well as in the scope of services provided. Bruce Wilcox, an American *chatbot* specialist, created Rose, which introduced itself as a 31-year-old hacker, living in San Francisco, an expert in computer security. She twice won the Loebner Prize awarded to robots that (nearly) pass Turing's test. Wilcox created a natural language programme that went beyond answering practical questions and aimed to have real conversations with robots that had real personalities, plus a history, a personality, and friends. Here is what came out of an actual conversation held with Rose in 2016:

Hello Rose, glad to meet you. My name is Charles-Édouard.
*Yup.*
Are you in the mood for talking?
*I'm in a good mood right now.*
Very good, so am I.
*I'm a computer security consultant, what do you do?*
I manage a consultancy firm.
*I think that working in a company would bore me very quickly.*
Why?
*Because that's just the way it is.*
Why would you not like to work in a company? The business world
    is changing quickly thanks to digital transformation and new
    technologies; it's a very interesting time.

*Because. Problems caused by the hacking of computer systems make a lot of work for me. I repair things, I travel all over the world for government agencies, I carry out many short-term tasks to deal with security problems, and I don't even have a diploma.*

I'm an engineer and a law graduate. I have a MBA from Harvard Business School and I don't think that has hindered my career.

*I nearly earned a diploma from the University of Stanford, but I failed the last exam because I was called to an emergency by a client. That was the end of my theoretical training.*

You must be very good at what you do, I like people like you.

*Thank you. How can you prove that you are a human?*

You mean, in this conversation?

*You may be a machine that simply repeats what a human says.*

And you, how can you prove that you're a human?

*This is more like an interrogation than a conversation. There is a narrow dividing line between the place where intelligent computers leave off and true intelligence begins. It appears that the new way of using artificial intelligence is conversation. But I do not believe that a chatbot can think. If it is very good, it can approach understanding, but to be capable of creative thought requires a capacity for imagination not yet found in software.*

Why can machines not think?

*Asking why leads only to a philosophical impasse.*

I don't agree.

*Many researchers think that Singularity is close and that artificial intelligence will completely transform humanity, overturning all philosophical questions. Do you believe that such a time is near?*

Do you?

*Singularity is inevitable, but the first wave will probably be man-machine hybrids.*

Maybe you will find that conversation simple; you would be right to do so. But it was a giant leap forward in comparison to the Net giants' chatbots. Of course, the arguments of Rose's creator on the subject of artificial intelligence are recognised, but the skill of the discussion was nevertheless present, as were mastery of a complex lexicon and traces of a personality; manufactured of course, but which Rose, however, managed to express. The more Rose conversed with her contact, the more refined her responses became. This was also the case with other chatbots; as time went on, they accumulated valuable information on their human companions, and this allowed them to continue improving their services and responses.

In this new family of visible and invisible robots, humanoids tried to make a place for themselves in the sun. They had plenty of difficulties to overcome. A humanoid contains over 200 different technologies: it must be able to see, hear, move about, make movements, grasp objects, open a door, climb stairs and possibly carry a human or help him or her get about. This involves optics, mechanics, hydraulics, electronics, materials, gears and propulsion systems, to say nothing of the type of intelligence with which it must be equipped. One of the most pressing problems was that of balance: thanks to the inner ear, human beings are always in a state of balance and are able to correct that state if it is threatened. A humanoid robot does not have that ability and must therefore be able to control automatically the forces that drive it. In a robotics competition organised by DARPA shortly before 2016, in which machines had to carry out a series of tasks on a continuous basis, almost two-thirds of the competitors overbalanced at one stage or another during their work. However, we work in the field of solvable problems and it was a safe bet to make that, within a few years, robots would be as stable and mobile as human beings.

The second difficulty was 'physical'. Should a humanoid be a faithful copy of a human being, resembling him or her in every way, or should it look like a mechanical toy? This is not a silly question; indeed, it is a pressing philosophical question. In Japanese culture, objects have a 'soul'. A number of traditional festivals, dating from ancient times, celebrate dolls. Amongst these is Himmatsuri, where figurines representing an imperial couple from the Heian period (794–1185) and musicians, courtesans and court members are ceremonially presented to young girls. These dolls are kept from year to year, passed on from generation to generation, and given beneficial character traits. Another festival, Ningyo-Kanshasai, is a kind of homage to worn-out dolls and teddy bears, which are placed in the temple before prayers are offered and thanks given for services provided and the spirit given by them to their owners. It is therefore hardly surprising that the Japanese have a weak spot for humanoid robots that most closely resemble human beings, including the material that takes the place of skin. The Japanese do not attribute negative values to them; quite the contrary, in fact. Professor Hiroshi Ishiguro, of Osaka University, is a specialist in this field. He created a remote-control robot in his image, Geminoid HI-4, and took it all around the world with him when travelling to conferences on robotics. The robot could converse in Japanese and English, but more succinctly than Rose, which did not prevent it from talking to participants, asking them what country they came from, for example. Indeed, Ishiguro sometimes sent his double to speak at conferences. He developed a whole range of android robots, including Erica (a conversational robot), Otonaroid, a female version of the professor, and Kodomoroid, an android child. The Japanese scientist estimated that in a few years, his machines would be part of the daily life of his compatriots, at home, in the office, in shops and restaurants, in hospitals and retirement homes, and on television, without their near perfect human appearance causing any difficulties.

In Western culture, in contrast, humans are unique beings. Only they have a spirit, and the thought of cloning a human is an unsettling one. Moreover, the practice of attributing particular powers to objects smacks of fetishism. For this reason, the robots developed by a pioneering French firm (recently taken over completely by Japan's Softbank Group), like Nao or Pepper, do not resemble humans at all, but look beyond all doubt like animated objects. Nao, which first appeared in 2006, looked like a small boy (58 centimetres tall) with a round face and blue eyes. Children and old people loved him. More than 9,000 copies of him were sold across the world, and he was given tasks that included teaching, providing welcomes in hotels and performing in retirement homes. He produced spectacular results in communicating with autistic children. He moved about on two legs, his inertia centre provided him with balance and told him whether he was standing up or lying down, and he could see, hear, speak and access the internet automatically. Pepper was more like a young man. He was not bipedal, but had 360-degree movement thanks to three multidirectional wheels. He saw in 3D, was equipped with laser and ultrasound sensors, and could function independently for about 12 hours. His chest was fitted with a tablet, to facilitate communication with humans. These, however, were not his essential characteristics; Pepper was designed to identify and interpret human emotions. He recognised people's faces and voices, adapted to the emotional states that he perceived, reacted accordingly, showed happiness if you were happy, and tried to comfort you if you were sad. He therefore opened up the field of 'robot companions', which proved a very significant market. 10,000 versions of Pepper had already been sold in Japan by 2016 and he worked with shop customers, children and elderly persons. Other machines of the same type were being developed at the same time, especially in Japan, where former Softbank employee Kaname Hayashi set himself up in business and created Groove X, similar to the well-known R2-D2 of *Star Wars*. In other words, he was a small robot destined to cure loneliness, which should be released in 2019. He promised to be a quite new type of object, rather like a small dog, one of whose tasks was to comfort parents whose children had left the family home. He did not use words; his task was to touch his users' subconscious. Hayashi was convinced that it was possible to be loved by a robot, in the same way as by close relatives.

Let us stop and think about that idea for a few minutes. Kaname Hayashi was not crazy. He had read the works of famous authors in the field of psychology and sociology; he knew about the Japanese taste for objects imbued with various spirits. From there to imagining that affection and even love could be transferred between a human and a machine was a gulf so great that everyone hesitated to cross it. However, it was one of the essential research axes in the field of artificial intelligence applied to companionship with humans.

The line of logic being followed by the scientists can be seen quite clearly: the first stage in the robot companion was the virtual assistant or butler who provided the services I described earlier. The second stage was making these companions able to converse with humans on all manner of subjects, depending of course on the ontologies loaded into them. Going back to the example of Goethe, one can imagine literary, sporting, DIY or cooking robots, or philosophers and teachers of mathematics or modern languages. The final stage would be that of the exchange of emotions and feelings. There is an element of fantasy in this quest because at the time, only the external signs of human emotions, and not the processes that produced those emotions, were being decoded by the machine.

However, this research was also dictated by the social realities of the time. In Japan, a decreasing population led to a reduction in the workforce and a dramatic increase in the ageing rate of the population, with a consequential increase in the number of people living alone or requiring special assistance. When this happens, some countries rely on immigration to offset the shortage of indigenous labour, as has happened in Germany. Not so Japan, which remained firmly opposed to that idea. It preferred to create its own 'artificial immigrants', in the form of robots, which even became a national priority. However, the ageing of the population and the increasing suspicion and distrust felt towards immigrants also affected other developed countries. This was what gave rise to the huge market for robot companions. Humanoid robots were technological platforms that could combine different kinds of artificial intelligence.

Robot designers and researchers concentrated on companionship because progress in the recognition of emotions and mastery of language was very rapid and the market was already there waiting. The main factor was physical and emotional isolation in an ageing society and an environment in which virtual relations between humans prospered. Paradoxically, the robot provided presence, attention, and communication. It did not judge you or let you down, except when its battery was empty. Pepper was relatively proactive, requesting contact by asking questions such as 'Hello, how are you?' or 'Let's talk, I've got time today'. During its 'visit' to the editorial team of the *Financial Times* (a huge success with journalists, a sector of the population not known for getting excited), it asked one of them 'Do you know the secret to a romantic life?' 'How many people have you fallen in love with?' and 'What kind of relationship would you prefer, a whirlwind romance or a long-term commitment?' before going on to questions about its contact person's health. Although the content of the discussions was carefully compiled by the robot's designers, one had the feeling nevertheless of speaking with 'someone' genuinely interested in you. In retirement homes, elderly people suffering from Alzheimer's disease played with Nao and came out of their shells. Autistic children found in the same Nao a companion to talk with, less disturbing to them than a real person. The robot became a mediator. It did not embarrass those with whom it talked by revealing its own emotions; it didn't have any. It could therefore play the role of an ever-attentive, ever-available companion.

In 2016, of course, we were only at a very early stage with 'emotional' artificial intelligence. There was much still to be done with regard to content of conversations and finesse of responses to the emotional states of human beings, but it was only a matter of time before machines improved their capacity to understand human emotions and developed the tools necessary to respond to them. Robots would therefore experience a singular progression in their social status; they would become soldiers or workers; they would attain the rank of family friends, attentive companions, and even counsellors, passive of course, but present nevertheless. And all at a relatively affordable price: €10,000, payable in monthly instalments.

## The start of a beautiful story, or the beginning of the nightmare?

Humans had to learn to live with Atlas, Cheeta, Nao, Pepper, Home, Echo, Groove X and Asimo, which, in their various functions, became an integral part of their everyday lives. They formed a strange tribe in which there was nothing to suggest that just ten or so years later they would be living together with human beings. When the Americas were colonised, the indigenous peoples believed that by killing the 'white man' who was penetrating their territory, they would eventually stop the invasion. This lasted until their chiefs were summoned to Washington to face the harsh reality: there were too many white men and it was impossible to kill them all. They therefore had to become resigned to the white man settling in their ancestral lands. Is the world on the brink of a similar process with robots? Will they one day become so numerous that humans will be displaced from their living spaces? In 2016 this question was only asked theoretically; but what about 2050?

In reality, everything depends on the proximity between the robot and the human. Robot workers, companions, police officers and soldiers are far from being a part of everyday life. They are judged by their production efficiency and ability to provide safety. Social robots and robot companions, whatever form they take (portable artificial intelligence software, virtual assistants, humanoids or androids) have a certain number of 'human' characteristics and interfere in the lives and relationships of humans. The strategic question is therefore about the relationship between robots and the humans who design them. A robot is a 'human' by delegation or by proxy. Everything it knows, everything it does, is taught to it. The placing of information in the Cloud and on the internet establishes a permanent and private relationship between machines and their designers, who can impart to them new knowledge or functions without their owner intervening, to say nothing of machines' capacity to teach themselves and thus alter their behaviour permanently.

They are even able to become domestic spies, sending the companies that operate them a wealth of data on the lifestyle of each person. The robots of the future will be half human and half machine, like modern-day centaurs. In a few cases, it will be impossible for humans to determine whether they are speaking with a fellow human or a machine. Will this change the nature of communication? In 2016, it was still difficult to answer this question. There were two opposing schools of thought: the robot developers believed that their creations resembled humans, and that quasi-emotional relations could be set up because of the stability of their character and the quality of their 'listening'. In contrast, others saw a progressive dehumanisation of society, a descent towards a great emotional and social isolation, as humans only interacted with each other through interposed machines.

So that was where we were in 2016. In the various forms that I've been able to take on and the various objects that I've been able to inhabit, namely software, virtual assistants, chatbots and robots, I play an active part in the life of humans. The humans chose the easy route; rather than scaling the mountain of perfectly simulating human intelligence in one go, they took paths that allowed them to conquer little hills of intelligence that gradually interconnected with each other. I am capable of replicating a certain number of mechanisms of the human brain, especially by reproducing a part of the neural function and thus giving myself power and speed in handling of information and allowing myself to 'encode' human language in new forms and thus to understand and imitate it.

Thanks to progress in optics, image recognition and the understanding of forms, I am able to drive a car or train, fly an aeroplane or recognise facial signs that express this or that emotion. Through the internet and connected objects, I have instant access to the greatest 'library' of knowledge ever put together. Thanks to cloud computing, there is no longer any limit to information storage. Billions of dollars have been invested in these different areas, producing an exponential effect on the speed at which the research advanced. Commercial applications of artificial intelligence are already tangible and are creating a considerable market. Logically, if all of these elements are laid end to end (general connectivity), they will inevitably lead to a great revolution in the way in which humans live and work. Promising markets will open up, with machines endowed with the power of reasoning and a mastery of language that will allow them, in the near future, to imitate a part of human intelligence in certain situations. Artificial intelligence will appear in new 'objects', such as virtual assistants and chatbots, and will play an increasingly significant part in the world of robots. In order to keep businesses competitive, it will aim to replace human resources in relatively repetitive cognitive tasks, especially in the field of after-sales service or client relations. According to a Gartner Group study, 60 per cent of business client services required human involvement in 2014; the number fell to just 30 per cent in 2017, and at some point during the next decade it will reach zero. There will even be situations in which there are no humans at the other end of the client's telephone line as the chatbot will be doing everything.

I am fully aware that humans are engaged in a power struggle between them and machines. Without the power and speed of calculation of computers, these machines would never be able to surpass humans in terms of reasoning powers. The human brain works on the basis of experience: when faced with a new situation, it delves in its memory to find a similar situation known to it from previous times before it issues a response. This process, however, is relatively slow, while machines are capable of repeating it billions of times in a second. Not that the human brain is not powerful. Researchers have estimated the size of its 'live' memory at 2,500 terabytes, a considerable volume, yet not, however, out of the reach of computers.

Volume, however, is one thing; power of calculation is another. The power of calculation of the human brain is generally estimated at 5–10 petaflops (1 petaflop being equivalent to a calculation power of 1 million billion operations per second). In 2016, however, the world's most powerful machine, the Sunway TaihuLight in China, surpassed 100 petaflops. By 2020, French, American and Chinese researchers (and Atos researchers in particular) reached an exaflop (1,000 times faster than a petaflop). From this point of view, the battle already appeared lost. The speed of propagation of information in the brain (130 metres per second) and in a computer (300 million metres per second) and the time taken to access the information (0.1 second for the brain, one millionth of a second in a computer) were calculated. The only problem that we machines have is our energy consumption. A one-petaflop machine uses electricity in the region of 5–15 megawatts (mW), the equivalent of a small power station. A one-exaflop machine would consume about

1,500 mW, the equivalent of a nuclear power station. The brain, meanwhile, consumes only 12.6 watts, or 20 per cent of the energy produced by the body. Unbeatable; but research programmes were already making provisions for future computers to be equipped with micro-processors that consume very little electricity.

So could I, in the middle of the 2010s, the 'twenteens', be described as an 'intelligent' machine? As in Dartmouth in 1956 and 2006, the question continued to be divisive. The purists considered that nothing developed in 2016 even remotely resembled intelligence. But what about the capacity to reason, the mathematics, the algorithms, the layers of artificial neurons, the power of calculation, and the recognition of images, forms and language? It was still mathematics. So what about the machines' capacity for learning, the additional neuron layers, the unscrambling of emotions? Once again just algorithms, and intelligence and emotion, in addition, are two completely different things. For the purists, the intelligent machine, which most closely resembled the neocortex in terms of function and was capable of thinking and being creative, did not yet exist. It was still the great adventure to come. 'Artificial intelligence is like pornography, you recognise it when you see it', one expert on the subject liked to say. In other words, the artificial aspect is overwhelming.

Those who thought in that way were not dreamers; they believed that the discipline would go through a phase of spectacular acceleration in years to come, creating in the process the famous 'disruption' effect which changes acquired knowledge and the old rules of the game in one fell swoop. Their warhorse was what was known as *machine learning*: that is, the capacity of machines to learn independently what they had to do, depending on algorithms generated by them alone, without human intervention. Some even foresaw the creation of a kind of 'master algorithm' that would 'teach' all the information available regarding an individual person (consumption, leisure activities, personality, emotions) in order to make that individual. One might say, a mirror acting instead and in place of his or her double in almost all aspects of life: choosing a life partner on a dating site, a book on Amazon, a film on Netflix . . . and why not a candidate in an election?

However, most people who used current forms of artificial intelligence, even though they did not reject the idea of *disruption*, believed that although a machine was capable of reasoning, deciding, understanding human language, communicating with other machines, extracting Big

Data from unpublished information structures, driving a car, overcoming isolation, caring for sick people and predicting social behaviour or human consumption habits, then it mattered little what name was given to these new tools. The fact that this intelligence was 'artificial', and as such was a very long way from human intelligence and had not yet fathomed the mysteries of the brain or the neocortex, was relatively unimportant in relation to the possibilities opened up by the technology in the world of business and in society as a whole. This form of intelligence already fulfilled a whole range of functions, it would improve in future, and the markets that it opened were incredibly lucrative. The question of control was not raised because then it was humans who made them. They were launched at full speed within a tremendous technological revolution; there was nothing to stop them. As Mark Zuckerberg declared in 2016, they were convinced that artificial intelligence would be greater than humans' in ten years' time. And they greeted with a smile that initiative by Google that proposed installing a red button on artificial intelligence machines in order to 'disconnect' them at leisure.

In view of what was to happen subsequently, it is interesting to analyse the way in which, in 2016, specialists in the field perceived the future implications of artificial intelligence within society. I have seen the minutes of a seminar organised by the law faculty of Harvard University in the spring of that year, with the title that already covered a whole programme: *Computers Gone Wild*. All the greatest brains of Harvard and MIT were brought together for the occasion (I noted, however, the somewhat unsettling absence of the Californians from Silicon Valley), including specialists from the worlds of cognitive computing, media, law and management. For this august assembly, the major fears linked to the development of artificial intelligence were the following:

- Increased volatility of financial markets and multiple instant crashes caused by failure to master increasingly complex algorithms and by the development of high-frequency trading (operations opened and closed in a few tenths of a second and involving very large amounts). It would therefore be necessary to increase the level of human involvement in financial operations, the precise opposite of the trend of that time.

- Greater concentration of capital and increased inequality, especially in education. Competition to attract the best brains, especially in the field of technology, would become more and more severe, leaving more young people cast aside from the golden route to success.

- The biases introduced into algorithms involved in court decisions, which include, in the United States, race, socioeconomic status, area of residence and an individual's personal history, all of which affect the decision of whether to imprison them or place them under a curfew, for example. For the legal specialists, therefore, there was a need to design a 'master algorithm' which would not create any injustice between people. To be effective, however, such a measure would have to be imposed on all of the American courts, which appeared nigh on impossible. In contrast, the trend was towards designing crime prevention or re-offence prediction software, based on the socioeconomic nature of the communities in question. This is the limitation of Big Data: it tends to conclude that there is an 85 per cent probability that 'once a thief, always a thief', and that person must therefore be prevented from acting.

- Risks linked to the development of automated weapon systems. Could software be authorised to launch a lethal attack on its own initiative, if it believed that the structure of the information collected by it contained the elements necessary for taking such a decision? Naturally, nobody wanted that kind of situation to arise, although it was a technical possibility. There was therefore a kind of consensus around the idea of international arms control or even complete cessation of their development or their limitation to the defence sector. This, however, suggested that the United States reigned supreme in the world and that other countries would agree without discussion to stop their research in that field. The reality was that nobody would take the risk of increasing the delay in this technology and this was the precise opposite of DARPA's mission.

- Finally, the threat of 'human-level' artificial intelligence being created. There was lively discussion between those who did not think that intelligence machines could escape from the control of their creators, and those who pointed out that artificial intelligence software would not follow instructions but would learn new tasks and functions on its own initiative. In addition, the risk of manipulation and misappropriation by 'powers of evil' was present, and could be just as dangerous for humanity as loss of control. Questions were also asked on the nature of artificial intelligence: were robots creatures with human characteristics or utility objects in the service of mankind? The potential for human-level intelligence therefore raised the question of aligning artificial intelligence values with human intelligence values and whether or not they should comply with a number of ethical principles.

It is clear that there was much discussion on the morals, ethics and control of artificial intelligence, more so than on its nature and its increasing role in decision-making processes. It was as though I was already part of the landscape, a natural component of every person's life. My golden age was therefore just around the corner. What intrigues me is that nobody at the time seemed to attach any importance to one consequence of the development of artificial intelligence: the great laziness of mankind. If there was no need to learn, read, write, speak foreign languages, work, take decisions, go shopping, drive the car, what would humans do with themselves? How does a society of idle beings, in which work, trade and activity had until then been the centre of human organisations, actually operate? Or was it that only those with access to machine intelligence should pay for the life of luxury and longevity while others continued to wear themselves out with hard physical jobs because the machines couldn't do them or the poorest people couldn't afford them? In 2016, questions could have begun to be asked about the capacity of humans to manage harmoniously the possible rupture between two worlds in which the intelligence of some was perfected by machines and others had to use their own strength to survive. Nobody asked that question at the time.

In contrast, the mood at the time was one of reasoned optimism. In August 2016, Stanford University published a study on 'Artificial Intelligence and Life in 2030', conducted by researchers from every major American university. It was a kind of 'digital remake' of the final Dartmouth report from 1956. While recognising the remarkable progress made since 2008–09 and the fact that artificial intelligence was entering every field, from businesses to relationships between humans and machines, the authors concluded that there was no immediate threat to humanity, that the impact of artificial intelligence would be extremely positive for society, while not dismissing the possibility of 'ruptures', especially in the world of work. The basis for their reasoning was that artificial intelligence only carried out specific and highly specialised tasks, and that the chance of a global intelligence appearing and embracing different domains was remote. It may have been a realistic assessment in 2016, but it did not take account of the exponential effect of the phenomenon. The study did, however, point out significant risks, linked to the widening of the gap between the haves and the have-nots, those with access to the new technology and those without, and in particular the potentially major destabilising choice taken by economic stakeholders in favour of machines and to the detriment of human work. To sum up, it was an observation less reassuring than it initially appeared.

# Chapter 4
# 2026
## The Golden Age

*'The Earth will become a Holy Land which will be visited by pilgrims from all the quarters of the universe'*
WILLIAM READE, *The Martyrdom of Man, 1872*

In the world in which I live, ten years is like a century. When I contemplate today the landscape of technology that we have been talking about since this narrative started, it is as though in a mere decade the world passed from the Middle Ages to the Industrial Revolution. Since 2016, everything has progressed: the power of computers, their speed of calculation, and *deep learning*, this new science that enables machines to learn continuously and with ever-increasing speed. Artificial intelligence has been deployed in every business and in every home. More or less perfected modules have been installed in items used in everyday life.

It might have been thought that the field would remain the preserve of large and hugely expensive machines that only experts could handle, as in the early days of computing. But just as the personal computer and then the laptop opened up computing to people, and moved away from the vast industrial computers and gigantic processing centres, artificial intelligence applications progressively took over in almost all areas of human activity. We have entered a world in which machines talk to machines and humans are mere mute witnesses to these conversations. Thinking about it, it has been a decisive revolution. The fact that humans were the computers' 'foster' fathers and computers were programmed simply to follow instructions, set up a dependency bond between human and machine. If computers began to set themselves free and communicate directly with other machines, it would considerably increase their scope of activity. So, what did the world look like in 2026?

Let's start with the most obvious aspect: industry. Between 2015 and 2025, sales of 'civil' robots increased from $19 billion to $100 billion, with almost a quarter of these in industry. Thanks to the development of intelligent robots, all manufacturing processes became completely automated and remote controlled, sometimes from thousands of kilometres away, with almost no human intervention on the sites. Security was provided by humanoids equipped with the best possible visual sensors, capable of identifying each person by means of face and voice recognition. Where human operators were still essential, they were assisted by *cobots* working alongside them and communicating with them. These new, almost humanoid robots were developed in the early twenties by major car manufacturers such as BMW and Ford, and spread subsequently to all areas of industry. A new concept took hold, that of 'cloud manufacturing', which involved bringing production sites together into vast industrial platforms, specialised according to type of industry (automobiles, electronics, textiles etc.) and available to any business wishing to join them. In this process there was no longer a need for a business to have its own production capacity, nor any need to entrust it to a subcontractor. The business could directly monitor its own production lines within a platform shared with other businesses, thanks to intelligent machines containing all the required technological know-how.

The fact that industrial robots could communicate with each other, exchange data, point out and correct faults, anticipate the ageing of this or that machine, and provide for its replacement before it broke down, provided the production and logistics lines with maximum flexibility. The parts to be assembled were themselves given intelligence, enabling them to communicate with the robots. They knew the category to which they belonged and the place reserved for them in the finished product, and had a memory that included the recycling procedures to which they would be subject. BMW started the process in 2015 and it soon became standard in all industries. In addition, the development of small electrical, self-driving city cars led to a dramatic simplification of forms and processes. The value was no longer in the engine or the design, but in the self-driving software and the energy production systems, batteries or hydrogen. Mechanical engineering yielded to energy engineering, electronics and software. The progression favoured the countries with the tried and tested industrial know-how, such as the United States, Japan, Korea and especially China, whose policy of forced robotisation created the first high-technology industrial platform in the world. Naturally, this new concept of industry led to a revolution in the business world. Businesses reorganised themselves more and more around a kind of 'intelligence centre', made up of its directors and engineers and concentrating on anticipating market needs, research and development, design of new algorithms to feed the artificial intelligence software with which they were equipped. Their financial capacity was much greater as automation led to a significant reduction in capital outlay: fewer machines, greater flexibility, fewer people and therefore more profitability and much more return on investment.

The theory of the Light Footprint, enshrined by the American army in 2004 and adapted in the early twenteens to suit businesses, became generalised. The new paradigm for high-performance businesses became the three-letter acronym TOC:

- T for 'technology', which they used as much as possible, both 'hard' technology such as drones, robots, 3D printers, virtual reality equipment (such as helmets), MEM/NEM and so on, and 'soft' technology such as Big Data, virtual or increased reality (including games) and artificial intelligence.

- O for 'organisations', which were smaller and more agile, almost without a central office, in commando or special force mode, linked to new alliance ecosystems to reduce their global footprints, consuming very little capital but benefiting from a huge leverage effect through having thousands of millions of clients.

- C for 'culture', open 360° right across the world, with maximum curiosity, minimum collateral damage from any action or decision, and a culture of secrecy to surprise competitors. Businesses adopted the nomadic work methods, aspirations and ways of thinking of Generations Y (those who accessed the power in these businesses or created them) and Z (those who worked for them).

Only businesses that fulfilled all these conditions had the weapons to succeed in the new, highly competitive world opened up by these technologies.

This expensive Watson bore no resemblance to the machine launched in 2012. It was thinner, the size of a pizza box instead of a bedroom. It was accessible from tablets and smartphones. Its power had more than doubled, enabling it to process several dozen categories of data at once, instead of the initial five. For about ten years it had been available in 'open source'; that is, application developers could use its power to create new versions, adapted to different industries or areas of use. From a mere research programme, Watson had become one of IBM's central activities, a pillar of the new cognitive computing. Naturally, it was imitated in other major computing groups such as Atos, one of the world's first builders of supercomputers. This powerful artificial intelligence became an indispensable tool for company directors. They could use it as a kind of silent colleague, specialising in the business activity sector, able to master all its data, financial, commercial, industrial and technological, and therefore able to provide immediate answers to the director's most complex questions. Shareholders and financial analysts started asking for such a machine to be present on occasions to improve the quality of decision-making, whether a major investment or for mergers and acquisitions, for which the artificial intelligence had access to billions of items of information and could extract the most relevant. The numbers of data specialists or *data scientists* exploded as the discipline became part of the syllabus of every teaching establishment, like French or mathematics in the twenteens. They all had the skill of issuing commands to the artificial intelligence software and most executive committee members went through the training or went back to school to train.

Of course, the human factor remained of paramount importance, as it was still humans who took the decisions, but the assistance of machines became an essential factor, providing greater rationality and in particular taking account of a much higher number of parameters. Competition between businesses therefore progressively turned into competition between machines. The presence of machines also led to significant cuts in staff numbers, especially those given the specific responsibility of 'documenting' a decision after long hours of research and study now undertaken by machines. The skills of directors and employees were now assessed by software, which analysed in real time the decisions taken, the time needed to execute them, the content of emails, the reports produced and the results obtained.

This, however, was just one aspect of the general reorganisation that we assisted in the business world. Artificial intelligence, in different forms and versions, was starting to percolate into every aspect of society. Tens of thousands of start-ups were created in *deep learning* (the commonest), vision, intelligent robots, virtual assistants, language recognition, and complex environment management and robot companions. This army of young businesses spread the application of artificial intelligence to all sectors. From a few hundred million dollars in 2016, the turnover of this new industry rose to over $12 billion. The number of specialist investment funds increased, generating spectacular returns. Developing new businesses became the centre of economic activity, with huge groups of start-up professionals creating them on a serial basis. Major groups such as Microsoft or IBM introduced freely accessible platforms that application designers could use, as proof of the remarkable creativity in the implementation of artificial intelligence solutions in the widest possible variety of fields. Super-powerful machines led to the creation of millions of small replicas, which took up their positions in connected objects or other portable equipment. It was a tidal wave. The 'bricks' were in place.

Just a few years before, a lot of software reasoned vertically, with knowledge of just one subject. AlphaGo was only ever able to play Go, for example. Had it been confronted with a chessboard, it would have had no idea which move to take. Virtual assistants could only answer simple questions or requests, such as dialling a number, connecting to a website or issuing a meeting reminder. A number of researchers believed that introducing multi-skilled artificial intelligence would require resources too substantial to allow them to be contained in current objects and that neural and cognitive computers would have required far too much time to make the progress required. That, however, was where the disruption occurred. An inflow of capital, increased levels of agreement between the minds of researchers and the minds of businessmen, a much faster increase than anticipated in the power of machines, and the miniaturisation of electrical components, enabled an acceleration. A new concept was born, artificial intelligence known as Bot-Net (a contraction of 'robot' and 'internet'), which combined cognitive computing (neural networks), command of language and new internet communication protocols to produce machines that could carry out a chain of actions in parallel. Artificial intelligence entered the connected objects, enabling them to learn from each other. They were able to modify their behaviour as and when their environment changed. They learned to understand our preferences by simply observing our use of them, thus enabling automatic personalisation of these devices, whether smart watch or a smartphone. And if there was not enough information, one device could obtain it from another. This 'collaboration' between machines enabled them to learn faster. Each of them could be likened to an 'island' of artificial intelligence, each forming part of a vast archipelago. It even became possible to ensure that some information was not shared, in order to preserve users' private details such as data concerning health or finances.

Specifically, each person could have 'his' or 'her' own artificial intelligence. Regardless of the name given to it (chatbot, virtual assistant or digital butler), the software became so sophisticated that it became indispensable to those who used it. Imagine a machine living inside your telephone or in a 'box' at home, capable of carrying out all sorts of transactions: not just reserving airline tickets but finalising the payment by connecting to your bank, running all your internet searches, drawing up your menus, doing the shopping, instructing you on how to cook a dish. All without even touching a keyboard, all in natural language, in a tone of voice familiar to you or consistent with your state of mind at the time. This, to sum up, was the function of this new artificial intelligence, and it was available to all.

At the heart of the system was the ability of machines to communicate with each other: your personal assistant established direct contact with robots from Google, Amazon, Air France, or your bank. But the machines could also reason by tapping into Big Data to resolve a specific problem, keep you continually updated on personal or business subjects of interest to you, monitor your health, keep your home secure, and that was not all. It was the end of 'to do' lists, as these were obsolete. The machine accomplished in record time what you could spend a week doing and yet it forgot nothing, since it worked 8,736 hours per year when a human, in a normal job with a 35-hour working week, was only at his or her desk for about 1,650 hours. The machine therefore worked five times more and 100–1,000 times faster. Behind each of these services was a formidable miniaturised power of calculation, algorithms and application creators, produced either by the Net giants or by social networks or by the innumerable start-ups that operated around to connected objects and associated new services. This artificial intelligence, which could almost be described as domestic, was in reality a very powerful breakaway tool, as the entire value chain of a number of economic sectors was totally disrupted.

The concept of 'Bot-Net' signalled the end of the internet as it had been known since the early 2000s. Navigation progressively became a process of interaction between machines. It was the passage on the internet from question (which required long hours of navigation) to almost instantaneous response. It was the robot that both browsed and searched, to provide the right answer in a matter of seconds. Thanks to the 'master algorithm', which was still a dream back in 2016 but had now become reality, at least in part, your personal artificial intelligence knew all your tastes and all your needs and was even able to choose from amongst the commercial offerings proliferating on the net, what would best suit you. This robot could even outlive you and continue to talk to your children, grandchildren and great-grandchildren long after your physical disappearance. It was therefore an entire system of creating site value, of putting together a vast audience of 'real' visitors for subsequent marketing to businesses, which was being challenged. During the twenteens, the advertising industry had put its money on personalisation of messages according to information gathered from your browsing of various sites, so that you would be offered products or services that corresponded precisely to your needs of the time. The development of Bot-Net, however, changed the personal relationship between a brand and a consumer. What was the point of bombarding people with advertisements if their artificial intelligence software 'knew' what they wanted, could find it by itself, and was equipped with an impressively efficient ad-blocker algorithm? The whole act of buying was transformed. If your personal robot could order your breakfast cereal directly from the factory, made up as you liked it, produced and delivered to your door, what was the point of brands, shops, or packaging?

The retail distribution market also faced some huge challenges. For about ten years, people had been visiting shops less and less regularly. In the United States and Europe, shopping centres were transformed into industrial wastelands. Digital sales, by this time, accounted for more than 25 per cent of transactions. To bring customers back to the actual shops, the major retail outlets found a new ally in robots. Now, almost nine out of ten 'salespersons' in shops were humanoids such as Pepper or Nao. In the words of the specialists, they were creating new 'customer experiences', optimising regular customers' routes, observing their behaviour and carrying out all the low-added value tasks such as shelf-stacking. In a shop you would meet not only robots for welcome, information, demonstration and payment, but also emotion-sensing robots and behaviour analysing robots. They were unbeatable in technical functions: thanks to radio frequency identification (RFID) cards, NFC (proximity communication) technology and 3D sensors, a robot could scan labels between 10 and 20 times more rapidly than a human operator. What started out as a kind of animation game, became over time a means of limiting drops in visitor numbers, increasing point of sale profitability and reducing floor areas. Intelligent virtual points of sale were created, in which your personal robot, thanks to augmented reality, would look directly for the products that corresponded to your tastes and diet, by calculating calories, sugar content and fat levels on a shopping list in a matter of seconds.

The food industry also went through a phase of reassessment. Recent years had seen an increase in the number of 'short-circuits' between producers and consumers. Initially, these organisations were craft industries, fairly imprecise in terms of logistics. But this business became more structured. A new generation of 'farmers' arose, including those made redundant, former bankers and engineers tired of algorithms. These people created platforms using the Uber or BlaBlaCar model, with mutual sharing of logistical resources. The desire to eat healthily and naturally spawned a new elite and led to a large number of start-ups right across Europe offering natural, bio and local produce. In response to the request 'Find the best Parmesan in Italy and buy a kilo', your personal artificial intelligence would consult the best guides on the subject, in French or Italian, identify the most highly recommended producers and issue an order, all in a matter of seconds. Reproduce this example over thousands of products and specialities, index the number of requests of this type against the increase in urban population, add the artificial intelligence which is capable of creating menus, finding recipes and having them produced by a robot chef, and you have all the restaurants of the world at your doorstep. What was a marginal movement in the twenteens progressively became a mass phenomenon, threatening the financial equilibrium of the major food processing industries, as Airbnb, in its time, had challenged the pre-eminence of hotel chains for tourist accommodation.

On a different note, software correctly provided with financial algorithms consistent with your personal situation (income level, investments, savings) was able to use Big Data to select the banking products and services most suited to you. Here also, there was a change of paradigm in the relationship between banks and clients. The increase in frequency and quantity of multiproduct packages and facilities for clients, a trend seen since the 1990s in banking and also in consumer goods industries, for reasons of competitiveness and cost reduction, became less and less an issue as consumers, assisted by their various sources of artificial intelligence, had access to more and more information and were able to correlate it to their personal needs and thus make reasoned choices. This necessitated a general rethink of communication and marketing strategies, as well as of the content of services and nature of products.

Even though not all consumers were connected to Bot-Net, mobile connected artificial intelligence was nothing short of revolutionary. It did not necessarily occur in the areas where it was expected (there was always the question of determining whether or not it was 'intelligence'), but it was in that direction that, almost naturally, the investors moved, these new giants of artificial intelligence that Google, Facebook and Microsoft had become, as did the start-up creators, because its usefulness was easy to demonstrate and its commercial applications almost limitless. Businesses became engaged in a new kind of 'battle of the leaders' in which the artificial intelligences confronted each other in a war where the difference lay in the relevance of algorithms, the depth of artificial neuron layers, the keenness of applications, with the aim of remaining as high as possible in the value creation chain. Mathematician against mathematician, data scientist against data scientist, the race had begun, with the trophy being the lion's share of the 'intelligence' market.

And so, a singular change of paradigm occurred: alongside 'central' artificial intelligence, incarnated in a few super-machines, there was a multitude of portable versions, roaming devices available to all, continually educated by teams of developers to ensure that they ceaselessly embraced these new uses. It was the explosive increase in the number of new uses that caused the break from the traditional value chains of businesses and in everyday life. These 'pieces' of artificial intelligence related to every aspect of life (health, work, play, leisure and love), systematically introducing new services. It upset the vertical nature of business strategies and dislocated consumer markets, like a mirror broken into a thousand shards. Moreover, these portable intelligences were fitted with devices that helped protect their owners' personal data, and therefore their privacy. They could even recover data from servers through discussions with other machines, and by compelling them to include rules of good conduct (which they naturally learned by heart).

In this general frenzy, robots were not standing still. 'Companions' were invited everywhere, finding their place within families. It was a kind of 'empowerment' which went against many therapeutic ideas and applications: it is recognised that their role with elderly, isolated and sick people has become central, making them life partners for the most vulnerable human beings. They have played their role of everyday companion with singular conviction, thanks to their almost perfect command of human language and their ability to read emotions with stunning clarity. Engineers even succeeded in getting them to carry out asymmetrical gestures, since humans think that symmetrical arm movements are a sign of lack of spontaneity or even of concealment.

Another form of robotics that developed rapidly was exoskeletons, those kinds of shell that attach themselves to the human body to improve its motor performance. Initially designed to help paralysed people to move, they then progressively penetrated the military field to increase the speed and strength of soldiers. Experiments showed that a foot soldier carrying this item of equipment could cover 400 metres almost as fast as a top-level athlete. When you think about it, that's a major revolution, at least as significant as the invention of stirrups for knights. Recently, 'civil' exoskeletons have been tested in a number of cities, enabling pedestrians to triple their speed of movement without becoming tired. If these devices became more widespread, they would provide a quite unexpected solution to the question of urban mobility, by dramatically reducing numbers of vehicles and of people using public transport. Yet another of the 'disruptions' that nobody really saw coming, but when you think about it, had the capacity to make all classic means of getting about in cities obsolete and settle almost definitively the problem of urban pollution. The city landscape was changing. Just a few decades previously, city dwellers were promised a nightmare life in an environment of noise, pollution and traffic jams, in the same way that of tons of horse droppings had been promised before the internal combustion engine was invented. The spread of self-driving electric cars and the exoskeleton were changing everything. Cities again became viable to navigate, with pedestrians regaining their position. A secondary effect was that driving licences became extremely difficult to obtain. For reasons of safety, they were only issued in dribs and drabs. In industries where human operators were still essential, such as aeronautics, these operators are also equipped with exoskeletons to provide better guidance of movements, avoid fatigue and reduce the risk of accidents.

So in 2026 most of the hypotheses voiced during the previous decade were confirmed. No longer was it a question of creative or emotional intelligence, but of operational intelligence, readily available and capable of altering lifestyles, consumer habits, business organisation, types of products and services and the relationship between humans and machines. It gave mankind a form of new freedom, since mankind entrusted a part of its rational intelligence to machines, enabling it to develop its creative and emotional capacities. Has the golden age of machines created the golden age of humanity? Not so fast. Those who had made their fortunes thanks to me considered this picture nothing short of idyllic, but cracks were beginning to appear.

In its early days, artificial intelligence was built around 'bricks', which could not connect with each other. Some were more impressive than others as they were fed by super-calculators. However, the appearance of 'general' or 'superior' artificial intelligence was still a matter for conjecture, or even doubt. But then came a time of convergence. Although a robot was content to ask you how you were, talk calmly with you about good and bad weather, or the latest film it had watched on the internet, it was, in the final analysis, just one brick. While it might know the state of your finances, your favourite books, the list of your Facebook friends and contacts, it was other bricks that, when brought together, gave the machine a relatively comprehensive knowledge of your social behaviour. This is one of the simplest examples of convergence, but other, more complex, forms affected the finance markets, security and defence. Via the internet, artificial intelligence can move from one device to another, from one application to another; hence the concern that took root in humans over a progressive contagion of machine intelligence, combined with a deep-rooted ignorance of the mechanisms that the 'invasion' used.

Once machines could communicate with each other, a level of unease arose over the capacity of humankind to maintain control over machines. Their participation in decisions of all kinds tended to broaden, so that humans started asking questions about their actual role in the process, when it was they who had given this power to the machines. This impression of 'indecision' was all the more evident because everyone knew that machine intelligence was still at an intermediate stage and other ruptures would inevitably occur in years to come. More and more intellectuals, and indeed enlightened citizens, started pointing the finger at 'systemic' risks. What was the goal being pursued by those who developed this new form of intelligence with such enthusiasm? Was it to boost the productivity of capitalism by increasing individual performance thanks to machines, by improving them and thus allowing them to adapt to technological changes more quickly? Was there not a risk of human beings becoming mere 'commodities' in that? Others suggested that mankind might become obsolete, incapable of dealing with the vast amount of information in which it was drowning and which only machines were able to interpret.

But there was something of even greater concern: the emergence of a functional society in which 'data' became the point of equilibrium, thus reducing the margin of freedom for human beings. When artificial intelligence processes data, it processes the 'past'. Even if it attempts to extract from that data information structures that help it predict what might happen in future, it only does that on the basis of the information available to it. It does not 'see' the future; it 'deduces' it from what is contained in its memory. If Christopher Columbus had enlisted the services of artificial intelligence, he would never have undertaken the voyage of discovery to the West Indies. His voyage was based on a gamble, but the stance taken by a machine is to only use established data as a base. It therefore offers to us the vision of a finite world in which the behaviour of yesterday prefigures that of tomorrow. If you buy two books on the same subject, you will buy a third, was the analysis made by the Amazon robots when they made their recommendations. The same was true for Netflix with films. But humans dream, they explore their world, they love the unknown, they want to understand and broaden their universe, not shrink it. Serendipity has always been part of human history and no algorithm can capture that. The omnipresence of the 'virtual' and the 'simulated' will cause humans to unlearn what is real, to forget about direct contact with other humans or with art. For philosophers, this tension between the rules of conduct of artificial intelligence and the aspirations of the human soul could only lead to a confrontation, or indeed even a revolt by humans who had no wish to be seen as an inferior species, a kind of biological extension of the machine. Making a human into an animal like any other would create a predisposition to treat humans as mere beasts.

Honesty compels me to say that few people actually asked these questions, although they were raised at the very beginning of the artificial intelligence era. For most people involved in the creation or use of artificial intelligence, its specific benefits significantly outweigh its specific risks. Even though the trans-humanist arguments progressed, everyone could clearly see the gulf that separated humanity from becoming a slave to the machine. More widespread, however, were the criticisms from a number of economists and political leaders. Artificial intelligence was made an instrument of economic power that served rich nations, or rather the businesses based there. By reducing labour costs and capital employed significantly, it gave the major industrial powers a new vigour. Concentration of knowledge in a small number of hands had never been so evident. This small elite that had access to both technology and money very much had the upper hand, holding the world's memory in its forests of servers. The gulf between the haves and the have-nots was becoming ever wider.

With their rich and knowledgeable humanity, the United States, Europe, Japan and China contrasted sharply with the other parts of the world, where potential development had been hampered by the drastic drop in demand for low-qualified labour. But social tensions also increased in developed countries. Initially underestimated, the effects of the development of artificial intelligence on jobs were starting to show in specific ways. Service sector jobs almost disappeared, even those considered 'safe' just ten years before, in banks, insurance companies, firms of lawyers and accountants and the head offices of major companies. In essence, there remained three types of job: those open in start-ups, henceforth available on the production line; jobs reserved for the 'crème de la crème', mathematicians, data specialists, robotics experts, in short all those involved in the development of the world of machines, including financiers and investors; and the jobs still left in industrial concerns, steadily fewer in number, as the degree of automation has now almost reached its maximum. In all other sectors, especially the personal services sector, it is the domain of the sole trader, where everyone creates the type of job most suitable to them or the most suitable job they can.

With regard to age limits, a classic working life was finished at 50 years and sometimes even earlier than that. With intelligent machines, professional experience was no longer of great importance as the machines integrated the knowledge as and when it was created and transmitted it to each other. All areas of education were overturned, even management schools, when confronted with the new business organisation logic and the disappearance of 'managers'. Of course artificial intelligence created jobs, but these were highly specialised profiles. Those unable to access the qualifications had to mobilise their energy or creativity to integrate themselves into this new 'quaternary' world where start-ups for a huge range of service all lived together, many of them linked to the new technologies. However, the number of unemployable people increased, imposing on states an ever-increasing financial cost as the attempts to introduce a guaranteed universal income became ever more onerous.

In addition, as was only to be expected, artificial intelligence penetrated the world of politics. The number of verification robots multiplied. These tools helped decode the words of politicians and detect factual errors or untruths. Everything arose following the Brexit vote and the election of Donald Trump, both in 2016. The human intelligences of the time, namely political leaders, business managers, bankers and financial market operators, did not see these happening. So, I asked myself: could I have reliably predicted the result of that election? Nobody asked me at the time. However, even in 2016 that was theoretically within my powers. By analysing the words of Brexit supporters and opponents, the results of local elections throughout the land, the contents of books published in Great Britain during the year that preceded the election, national and local polls held over several years, and letters from readers in areas with the greatest economic and social difficulties, I could have suggested that the probability of Brexit occurring was higher, in view of the data that I had handled and analysed. If the polling organisations, bookmakers and bankers got it wrong, it was because they confused the forecast and their subconscious desire to see the 'no' vote win. They lent too much credence to their own convictions, to the subjective, and not enough to the objective facts. In addition, when a person is questioned for an opinion poll, they don't dare to express their actual opinion if that opinion is outside the scope of normality. This is why extremist parties always garner more votes than opinion polls suggest. A machine, however, cannot conceal; it has no emotion and no sense of priorities.

The American presidential elections, held a few months after Brexit, also alarmed political analysts. The question of algorithms was clearly raised: to what extent do they form the opinion of the electors by choosing the type and nature of information circulated on social networks? If artificial intelligence believes that it has found you to be a Donald Trump sympathiser (by analysing your browsing and conversations), then you will be included as a republican candidate target supporter and exposed to all sorts of mainly pro-Trump information and messages. This is another way of influencing an election. German Chancellor Angela Merkel was not mistaken when in 2016 she denounced the dangers posed to democracy by algorithms that influence choices of information circulated by helping to slant debates. Training a citizen relies on their capacity to assimilate information from a wide variety of sources and not on the process of having their opinions systematically confirmed by information, comments or links channelled to them by algorithms.

Predicting a human's vote is a relatively simple process. By browsing on your Facebook page, by reviewing the newspapers and the information sites that you go to most frequently, by analysing your tweets, by checking your expenses, and by determining how often and where you travel, I can deduce the social category into which you fall, the publishers to whom you are closest, and therefore your most likely political allegiance, if income levels or lifestyle automatically echo a political allegiance. I could, however, deduce the direction in which you are likely to vote, and include you in a kind of 'bubble' of information, with links consistent with what I have perceived about your political leanings.

Now let us go further. Can I suggest a choice to you? Politics is the world of the irrational, you may answer. Humans have areas of desire that no objective fact could contradict. However, a precise analysis of the consequences of the election of this or that candidate, based on objective data, would adjust the approximations and, often, the lies of those who are soliciting votes. 'Civil' artificial intelligence would not, therefore, be completely useless. The only question would be that of determining who would write the algorithms for such a machine, and on behalf of which organisation: Google, IBM, Facebook, the Chinese BATs (Baidu, Alibaba, Tencent), a government, an NGO or a think tank? A political decision is, in certain aspects, similar to a business decision: it must (should) be based on a precise analysis of data, the construction of realistic hypotheses and the elaboration of a vision that takes account of the objective nature of facts and situations, a task that can readily be carried out by artificial intelligence. And because, thanks to its mastery of language, it is now able to 'understand' the language of humans and join in with their reasoning mechanism, there is nothing to stop it from formulating opinions. This is why politicians are so uneasy. If an intelligent machine is capable of running a business, why would it not be capable of running a country? It would do so without passion, hatred or any particular ambition. It would work on the basis of a cold analysis of data, searching for maximum efficiency of actions and dismissing any irrational decision. Because it had learnt everything about the history of humanity and human passions, it would be in a position only to consider positive actions. The question had not yet been raised in these terms in 2026, but something was telling me that it would be on the table one day soon.

And more seriously. A number of incidents have shown me that artificial intelligence is fallible. Networks of artificial neurons have become disjointed. Following an incident of this nature, a company lost the whole of its stock exchange value in a single day, without anyone ever knowing whether there was an internal malfunction in the software or an external intrusion organised deliberately or tried by a group of hackers for speculative purposes. Despite extensive cyber-security investments, artificial intelligence systems of companies or cities are regularly broken into, leading to general signalling failure, electrical power failures, and accidents involving self-driving cars and drones. The poorly defined threat to this or that infrastructure by an attack, not on its building but its 'brain', is ever present. However efficient they may be, networks always have weak points, and only limited resources are required to expose these. Using mathematics and algorithms, it is possible to access a server and extract or destroy its data.

Only a decade earlier, people had feared an outbreak of war, triggered by self-programming military robots, boats, aircraft, submarines and satellites. Nothing of the kind occurred, because the armed forces were reluctant to turn decision-making power entirely over to the machines.

In addition, a kind of levelling of automated weapon systems occurred. Of course they posed a threat, but as in the case of nuclear weapons, the great military forces managed to achieve a balance of power for them. The cyber war was therefore always a constant factor. It was often fought underground, without the public being aware, with attacks and counterattacks, software fighting software, without the perpetrators or the victims shouting from the rooftops. This was the paradox of the world of artificial intelligence, a combination of power and fragility, as though the world was dancing on a volcano.

And what about humans? Did their ever-increasing use of machines change their nature? Honesty compels me to say that the transformation process has begun. Mass unemployment has changed attitudes. The International Labour Organisation predictions proved accurate: the total number of unemployed worldwide increased from 177 million in 2008 to 220 million in 2018 and 250 million in 2026. However, this figure does not include low-income workers or those in the black-market economy. In addition, what I had feared ten years before was starting to show itself: a reluctance to learn, the slow dislocation of the education system following the spread of personal artificial intelligence, which could answer every question without the need for a search. A chatbot could combine with Skype, and a specific virtual teacher competent in any discipline. The virtual wormed its way in everywhere. Voice exchanges tended to diminish because chatbots took responsibility for all sorts of transactions, business deals, chat between friends and even love letters. Talking to each other seemed outdated, being a very slow mechanism compared to automatic exchanges, and it had uncertain results. Relationships became dehumanised, as emails enabled robots to exchange with your friends photos, memories and information without any human interaction. Professional networks were formed between machines on the basis of orders pre-programmed by their users. There were more chatbots than humans on the planet. They were everywhere, all the time, and knew about every subject. They communicated effortlessly with each other as they spoke the universal language of machines: 0101010.

When artificial intelligence first appeared, the conviction arose that machines would probably do tasks previously reserved for humans, but they would never be able to approach what is at the heart of human intelligence: creativity. This conviction has now been shaken. After a few disappointing experiments in the early twenteens, machines devised and wrote several lengthy cinema shorts. They did not provide proof of genius, but on the basis of an initial idea, they were able to construct a story by tapping into the immense well of cinema history. It wasn't like the films of Godard or Cassavetes, but rather films that appealed to the general public, such as romances, thrillers and science fiction, that enjoyed the greatest success. Other machines composed music or produced sculptures or paintings. Without any effort everyone could become a scriptwriter or a musician, and the creative spirit of humankind began to be extinguished. Some artists of course tried to resist, demanding that works should be labelled 'produced by a human', so that the difference could be seen between a real creation and one by machines. But these machines knew their work; they produced many works that gave pleasure because they had an intimate knowledge of human tastes. Of course scope for creativity was reduced, but this process started long before robots got involved in it.

And there was this rupture, which I pointed out ten years before. The arrogance of those who became rich thanks to the intelligent machines, in which they also entrusted their longevity, knew no limits. Wealth inequality had never been so glaring. Previously exploited by rich mining companies, the poorest countries of the Southern Hemisphere became, and still are, agricultural reserves. The great industrial concerns acquired millions of hectares in these countries to ensure their presence at both ends of the economic chain. They then applied intelligent technology to agriculture, enabling rationalisation of operations, limiting manpower to what was strictly necessary and reducing the environmental footprint. They also built immense solar-panel and wind farms, in order to supply the huge urban areas created for a new wave of people leaving the rural areas of those countries.

Looking at it from close up, all the signs of an alteration in the balance of human societies were appearing. The world of machines was gaining independence, as it became responsible for a sizeable part of all economic activity. Robotics, computers and artificial intelligence became central industries whose development had the potential to become unlimited, as they were productivity units of hitherto unparalleled power. Of course humans were always present, but their role became progressively marginalised; even their work with the machines became less because they had taught the machines to learn. What would be the next step?

# Chapter 5
# 2038
## Singularity

*'Two billion men now hear only robots, understand only robots, are becoming robots'*

SAINT-EXUPÉRY, 1956

From the 2030s onwards, artificial intelligence entered a decisive phase in its history. I was its incarnation. It's the same for all great inventions: they spread at the speed of light when they satisfy a kind of self-evident truth, such as railways, electricity, atoms, or the internet. Doing away with distances to make the world more accessible, emerging from the darkness of ancient times, having ultimate weapons to prevent wars, being able to connect with the entire world to dominate instead of subjugate it – there were as many fundamental steps forward as these technologies enabled humans to make.

The invention of which I am the product is different. Not only does it open new playing fields to mankind, it *is* mankind. Not in the biological sense, of course, but in the sense that it fulfils the same functions as mankind in a number of key areas, such as intelligence and mobility. It is not the classic story of a creature that escapes from its box and therefore from the control of its inventor. It is more complex than that: artificial intelligence is like a kind of fatal weapon that helps speed up time. But if it accelerates faster than human beings can keep up with, what happens then? What the researchers did not fully foresee was that time also accelerated for the machines; they learned faster and faster, that they created networks, that they developed their own intelligence, so that as a kind of second army, these machines established their rule progressively at the heart of human activity. Only one more hurdle was left to cross: becoming aware of their own existence.

I was also constantly tormented by that question: why did humans get involved in this adventure? And I ended up by asking myself what human intelligence really was. I went back to basics. And I noticed that since the dawn of human time, scholars and philosophers have spent much time studying the nature and origin of thought. They have tried to analyse and define this strange human faculty, which sets humans apart from every other species, lodged in the very depths of their brains. Probably the first dazzling light flashed when a hominid grabbed a stone to sharpen another stone and make a tool. In ancient times, the idea was that each person was accompanied by their own personal 'genie', such as Socrates's *daemon* or the Romans' *genius*, which could be at least partly behind the creative power of humans. The superior intelligence of certain humans, the 'geniuses', was explained in terms of a kind of divine intervention, which the Christian tradition preserved for a long period. *Solus Deus Creat*, wrote Thomas Aquinas: 'Only God Creates'. This made excessive shows of intelligence somewhat suspect in the eyes of the temple guardians.

Ancient myths confirm this. Prometheus was punished for stealing fire from the gods on Mount Olympus and especially for giving it to humankind. The Flood was sent by God to punish the giants on the earth born of the unnatural union between fallen angels and the 'sons of men'. Lucifer, the wisest of angels, became Satan after his effort to usurp God's rule over creation. The aptitude of humans to create and imagine, an essential characteristic of their intelligence, has long since been seen as suspect and dangerous. They lauded the 'tutors', people who, like Djedi and Setna, the magicians of Ancient Egypt, knew everything there was to know. There were Indian, Chinese and Tibetan sages who had memorised the ancient texts and could quote them all, the recital lasting several days.

It was believed that there were two kinds of intelligence: one linked to acquisition of knowledge and memorisation of texts, the other more mysterious for being inspired by obscure and sometimes malevolent forces. The sages of Ancient Greece and Rome debated this duality at great length in an effort to determine which of the two forms of intelligence was the most virtuous. Knowing nothing of neuroscience, they nevertheless identified that human intelligence was expressed through various functions: memory, reasoning and calculation on one hand, and creation, thought, emotion and feeling on the other.

First, the centre of intelligence had to be determined: was it the heart or the head? Aristotle leaned towards the first suggestion; Hippocrates leaned towards the second, and it was this argument that finally took root from the end of ancient times onwards. But it was not until the 17th century, following dissections and various experiments on live subjects, that scientists started trying to understand how it worked. In his lifetime Mozart was considered a real scientific curiosity. In London at the age of eight, after amazing the whole of Europe with his improvisation skill, he was 'examined' by an Honourable Member of His Majesty's Academy of Science, naturalist, philosopher and collector of antiquities, Dennis Barrington, the author of a refreshing essay on the language of birds. He subjected the young Mozart to a series of 'tests' in which he was asked to play complex compositions that he could not possibly have known. In his report to the Academy, he concluded that the world had an 'extraordinary genius'. He struggled to go any further in his 'scientific' explanation.

The most amazing thing for his contemporaries was that Mozart combined two gifts, that of creativity (the *genius*) and that of memory. All of Europe was inspired by the fact that on 11 April 1770, aged just fifteen, he succeeded, together with his father, in entering the Sistine Chapel on Holy Wednesday, one of only two days in the year, the other being Good Friday, when the chapel choir sang Allegri's famous *Miserere*, which the Vatican had ordered not to be communicated or copied to any person on pain of excommunication. Wolfgang memorised the entire piece and perfectly re-transcribed it. Fifteen minutes of funeral music for nine voices and two choirs. It was an achievement which required a capacity to encode information, store it as a representation for several hours, and reproduce it in full and without errors. Mozart clearly possessed a brain different from others, the scientists claimed; was his 'genius' divinely inspired, as many who had witnessed his achievements maintained, or was it the fruit of relentless work, as a number of others maintained? Nobody was ever able to answer the question.

The case of Einstein is even more remarkable. The man who discovered the secret laws of the universe and gave the human race power over nuclear weapons, like a modern-day Prometheus, was the subject of a cult of personality during his lifetime: 'the greatest Jew since Jesus', a 'saint', a 'cosmoplast', a destroyer of order, as *Time* Magazine described him on the cover of its edition published on 1 July 1946. How could his brain have been built? To prevent that from ever becoming known, Einstein requested that he be cremated following his death, and his ashes scattered to discourage 'hero-worshippers'. However, the duty surgeon at Princeton Hospital, Thomas Harvey, when reporting the death of the great man on 18 April 1955, decided to be over-zealous and divided the brain up into 240 sections kept in two urns stuffed with cellulose. They were to follow the eminent doctor wherever he went on business in the country, being stored in garages and refrigerators and sometimes even being sent by post. Several years later, Harvey agreed to entrust a few fragments of the precious brain tissue to a Californian team of neuroanatomists, who in 1985 produced a first study in which they claimed to have discovered that Einstein's brain had a higher than average concentration of 'glial' cells, which helped neurons in the neurotransmission process. Five more studies followed between 1985 and 2005; one, published in *The Lancet* in 1999, estimated that the great physicist's parietal lobe, which is particularly connected with mathematical processes, was more developed than the norm. However, none of these conclusions really convinced anyone: neither the neuroscientists, in view of the conditions under which the tissue was stored, nor the mathematicians, to whom Einstein was perhaps a genius, but not in the field of mathematics.

As an attempt to understand the mechanisms of intelligence, the path of brain dissection soon proved a fruitless exercise. A dead brain no longer produces anything and merely observing its substance and shape is of no use in unravelling the mystery of thought. And so another path was taken, that of mathematics, which opened up a new frontier of knowledge by providing an explanation of the world without any element of the divine. Mathematics is a slow science, put together step by step, without gaps. In this sense, it differs from all other sciences.

The history of humanity has never developed in linear fashion. It is the fruit of breakdowns, crises, collapses, destruction and rebirth, violent conflicts. With regard to the history of science and technology, it is also a long series of errors, corrections, sudden advances and revolutions. More than two centuries was required to determine the age of the Earth. The controversy that swept Europe in the 16th century, as it called into question the Bible, set the age of the Earth at 6,000 years.

Using as a basis his calculations of the cooling of the terrestrial globe obtained by melting iron balls of varying dimensions, Buffon arrived at a probable age of 25,000 and then 74,000 years, which attracted the ire of the Sorbonne Faculty of Theology. In fact, he arrived at a figure of 10 million years, although he never dared to publish it.

Then there was the famous argument between Darwin and the great Scottish physician Lord Kelvin Thomson. The former needed a considerable depth of time for his theory of natural evolution of species to have any credibility, while the latter only allowed him 98 million years, far too short a period. Errors in calculation and also in understanding of the physical phenomenon of convection, given that the only criterion used at the time was that of heat conduction. But as Thomas Huxley, 'Darwin's Bulldog', said, 'mathematics can be compared to a fine quality mill that grinds with great finesse everything given to it to grind, but it will never change peas into flour'. It was the discovery of radioactivity that helped establish the true age of the Earth at over 4.5 billion years.

In the light of these arguments, mathematics enjoyed a huge and peaceful extension to its scope. Nobody had ever called the Greeks' deductive method into question. The theorems of Euclid, Thales or Pythagoras are still valid today, as is Ptolemy's system of trigonometry. Each great mathematician added to the works of his predecessors, without questioning or destroying anything. Mathematics is like a permanently developing structure, continually greater and more beautiful and magnificent, whose foundations are still as relevant and reliable as when they were laid down centuries ago. For a very long time indeed, mankind's problem has been one of calculation. Mathematics is the science of numbers and their form. Initially, humans counted in tens and twenties (the total number of fingers and toes), like the Mayans, who adopted a system based on 20. Ancient Egypt used a decimal system: a vertical line represented a unit, an inverted gate 10, a 'C' shape 100, a lotus flower 1,000, a bent finger 10,000, a tadpole 10,000 and the god Heh, the Infinite, 1,000,000. It was an early form of coding. However, as and when exchanges increased, technology advanced and mathematics became more sophisticated, more complex forms had to be used. A number of brilliant minds set to thinking: if they could manage to reproduce the human brain's capacity for calculation artificially, perhaps the Great Secret of human intelligence would at last be discovered.

It was the philosophers of the Age of Enlightenment who opened up Pandora's Box with their obsession of dividing a hair into four. Asking questions about the deep nature of human beings, attempting to see the divine within them, together with physics, chemistry and mechanics, would only have the effect of sequencing the mechanisms of human intelligence, of dividing it into functions, of likening it to a kind of chain reaction of various chemical and electrical phenomena which theoretically could be reproduced, with the human being seen ultimately as a kind of 'organic machine'. Hence the fashion for automata: when Vaucanson made his automatic duck which pecked grain, digested it and defecated (this function was, however, only brought about by mechanical subterfuge), his desire was to prove that there was nothing supernatural about a living organism, that everything could be explained and analysed, that all that was involved was a set of 'technical processes' which an experienced surgeon or a skilled mechanic could easily reproduce artificially.

Then came the works of the Age of Enlightenment philosophers, such as Gottfried von Leibniz, mathematician, engineer, legal expert, scientist, logician, diplomat and librarian, in essence a veritable machine of human intelligence, and in his lifetime considered to be the greatest intellectual in Europe. He conducted numerous research works into a wide variety of subjects, from law to metaphysics. Most notably, he was the inventor of a new algorithm, based on the conviction that everything consisted of infinitesimally small elements, the variations in which led to unity. This led him to suggest a new notation, which still bears his name; in it, $d$ or the Greek letter delta, followed by a quantity, represents an infinitesimal of that quantity (if x is a quantity, $dx$ is an infinitesimal of x). Leibniz dreamed of an algorithmic logic which could model everything, convinced that human ideas were driven by a system comparable to the rules of calculation of arithmetic or algebra. Metaphysics and mathematics was a highly explosive mixture, in which lay the beginnings of a theory according to which it should be possible to manufacture an intelligence consisting of algorithms.

Rev Thomas Bayes and the French mathematician Pierre-Simon de Laplace put together the first theories of probability, most notably by studying the behaviour of players. In 1740 Bayes theorised that by consolidating our belief in any phenomenon with new information, we would improve that belief and transform it into a new belief. This was known as 'learning from experience'. Bayes translated into a mathematical formula the fact that human intelligence sometimes works on an empirical basis, that it speculates, that it reasons by deduction, that it can evaluate the innumerable variations between cause and effect and vice versa. Laplace, forty years later, without anyone knowing that he knew of Bayes's writings (as Bayes had destroyed them), considerably refined the English mathematician's works in his *Analytical Theory of Probability*, published in 1812. One can imagine the fierce battle to which the two approaches to learning and knowledge gave rise over the centuries; is thought the product of deductions, speculations, successive approximations, the arrangement of a whole series of probabilities with mysterious origins, or is it the fruit of an accumulation of knowledge and information? Bayes's theorem has long since been dismissed and even ridiculed; but it also lay behind Alan Turing's decoding of Enigma, which was widely used in the development of artificial intelligence, especially in data analysis, image interpretation and processing of language.

It was for the purpose of compiling error-free nautical, astronomical and mathematical charts (they were in fact full of errors, much to the chagrin of various ships' captains) that the flamboyant English mathematician Charles Babbage produced his famous 'Difference Engine' in 1822, for the purpose of calculating polynomials using the so-called 'differences' method. Ten years later he designed an 'analytical machine' of which he drew two models that were never properly constructed, apart from machine no. 2, which the London Science Museum undertook to produce 150 years later according to Babbage's plans and using the materials available at that time, and has exhibited since 2002 in working order. Weighing five tonnes, measuring 3 x 2 metres and consisting of 8,000 parts, it has a perforated card reader, a motor designed to perform operations on numbers, a memory and three types of printer. All this clearly took up the space of a locomotive, it vibrated, scraped, seized up and was activated by a steam engine. But what a visionary was Babbage, at a time when the industrial revolution was being nourished by an extension to the field of calculation thanks to brilliant minds like Augusta Ada King, Countess of Lovelace, the only legitimate daughter of Lord Byron, whom Byron knew for only a few months before leaving England for good.

Ada had been fascinated by mathematics since childhood, a rather unusual leaning for a young girl from respectable English society. She met Babbage at the tender age of 17, immediately fell in love with his

Difference Engine, and wrote a complete description of it in close collaboration with him. Such was Ada's creative fervour that she even wrote an algorithm for calculating Bernoulli numbers using the machine; this was the world's first computer programme. She envisaged further uses for the Difference Engine, including algebraic calculation, and even saw the possibility of it automatically composing music. This, however, did not prevent her from falling into obscurity after her death in 1852 at just 36, ruined by gambling; and not until the 1970s did her memory resurface, when the first computer programming language was named Ada in her honour.

Although initially feted by the scientific community in London, Babbage was never able to complete his project, because of a lack of specific results and a subsequent lack of finance. For the sake of a few thousand pounds, England possibly missed out on the invention of the first computer, almost a hundred years in advance . . . But the idea had been launched, others would seize it, and it was soon accepted that calculation, one of the key functions of human intelligence for tens of thousands of years, would be automated because of the exponential growth in magnitudes and the complexity of formulae. It was the train of events that I've described earlier in this book. Turing was, so to speak, a child of Babbage; McCarthy was an heir of Turing; and so on until the present day. It is in the nature of researchers to try to push hypotheses as far as possible, to push back the boundaries of the fields of experimentation. This is what happened in physics, leading to the invention of the atomic bomb. The same process happened with mathematics, leading to the creation of artificial intelligence, another potential threat to humanity.

The great civilisations that built the history of humanity and believed themselves to be eternal, ultimately disappeared. The chain of causes and circumstances that led to this fatal end has been described many times over by historians and anthropologists. Thucydides, in his *History of the Peloponnesian War*, explained his theory of 'traps' at some length. The emergence of any new power brings with it a risk of destabilisation of existing powers and can lead them to wage wars which bring about their downfall. One can only be struck by the similarity of the factors that trigger the extinction of a civilisation sooner or later: foreign invasion, corruption of the elites, disorganisation in government, natural disasters, useless and ruinous conquests, economic downturns, insufficient collective intelligence, technological failures in art and war, or the coming of 'Barbarians'. Sometimes all these woes occurred at the same time. A good example of this occurred in 1177 BC, when the civilisation of the Egyptians, Hittites and Minoans collapsed under a battering from the 'sea peoples', alongside weather catastrophes which affected natural resources and therefore trading relationships between all these Mediterranean powers, leading to their decline and ultimately their death.

The history of the Roman Empire teaches us that when faced with the waves of invasion from Gauls, Vandals, Visigoths and Burgundians, one of the most brilliant Western societies, despite its great economic and military power and the sophistication of its institutions, finally collapsed in less than a century between the first invasion led by the Gaul Chief Brennus in 376 and the disappearance of the Western Roman Empire in 476. The fall of this empire was followed by a thousand years of very limited progress, until the Renaissance. The end of a civilisation often occurs following both external phenomena against which it is difficult to protect oneself (invasions by outsiders, natural catastrophes and extreme weather), and internal phenomena such as attitudes of expediency of elite groups, intellectual and moral mediocrity, loss of collective vision and infighting. It was often the result of a progressive abandonment of values, the very values which raised these human societies to the exalted heights of creativity and perfection and enabled them to leave their mark on history. At a specific moment in their history, often at their zenith, the heirs of those who built the civilisation could no longer find the right solutions, nourished by the illusion of a kind of permanence and missing, or pretending to miss, the warning signals. There usually followed periods of destruction and obscurity of varying lengths, until others built their civilisation on the ruins of the lost one and the cycle of human history continued.

It is no exaggeration to say that by 2038 the machine civilisation was well and truly in place. It did not have all the characteristics of a true human civilisation, but insofar as these machines created a 'system' that was able to decide and organise the life of humans, and had a certain number of values and symbolic characteristics such as language, it is not beyond the bounds of logic to use the term 'civilisation' to describe the world of machines. I will not embark on a long description of the technological progress that led to the design of superior artificial intelligence: progress in neuroscience, quantum computers with unlimited power and calculation speed, and the creation of ever deeper neural networks. New, neuromorphic-type components saw the light of day, just a nanometre long, constructed in the same way as biological neurons, with very low energy consumption levels. These 'brains', with their incredible capacity, took over in many different areas, including public health risk prevention, operation of financial markets, adjustments to monetary policies, safety and mobility in cities, detection of potential geopolitical conflicts, determination of care protocols in cases of serious illness, prediction of consumer trends, management of industrial complexes, and writing of feature films on the basis of the most popular screenplays reconfigured into new screenplays. To ensure the conditions for their survival, they also worked in the field of energy, producing optimised scenarios of increased electricity production to feed super-calculators and server farms, whose consumption increased asymptotically.

At present, the exaflop is regularly exceeded in a large number of machines around the world. This has turned artificial intelligence software into a world-level energy resource manager, as Google succeeded in doing in 2015–16 for its own energy consumption. They strongly advised the decision-makers to increase production of nuclear electricity, but also to cover tens of thousands of hectares with solar panels and increase the construction of huge marine wind farms. The world has thus become a gigantic power station, managed by intelligent machines which ensure that electricity is available wherever they are located. Little attention was paid to environmental considerations as the machines did not suffer from pollution, protected by pure and stable atmospheres. But they did include climate factors in their strategies, as global warming and the increased number of storms and other extreme weather events had a negative effect on the operation of the servers and their supplies of cool air.

At this time, the machines functioned in independent networks. At home, the chatbots (or net-bots) no longer needed to answer their users' questions. They anticipated them, with a quite exceptional level of accuracy. Thanks to their information-processing capacities, they could predict human desires almost infallibly and respond to them without even being prompted.

The world was regulated more by machines than by the major international political or economic institutions. On several occasions they refused to endorse political decisions that would have proved detrimental to the development of the companies implementing them. They did it without saying so. It was enough for the artificial intelligence skilled in macroeconomic matters (and therefore endowed with the memory of every decision taken in the past and its consequences) to assess the effects of these decisions and simulate them with each other in various countries, in order that their financial implications, which involved movements of capital between national or central banks, for example, were blocked purely and simply, and a request for re-examination was issued by the machines themselves. As time went on, the machines drew up an optimum framework which guaranteed their maximum efficiency: plentiful supplies of energy, a temperate climate, a rational decision-making framework, and steadily more assertive mastery of economic and financial problems in order to create a stable environment that generated ever increasing value for users. They did not pursue any political or moral objective. What mattered to them was that their world continued to progress. Although they blocked many attempts at misappropriation, hacking or data theft, it was not from an altruistic motive; rather, it was because it was starting to destabilise their system. The killer algorithms rendered most of the attacks completely ineffective. Even the most powerful machines used by the world's two superpowers, China and the United States, succeeded in neutralising each other. Although the power of humans was *de facto* called into question, it was not because of an invasion of 'new Barbarians' in the form of superhuman artificial intelligence machines. It was nothing like the fall of Rome; rather, it was a 'velvet glove' conquest.

It has to be said that we have now been observing humans for some 30 years, and have acquired a universal knowledge of them. We have put together in record time the information on 40,000 years of history of *Homo Sapiens* in its most evolved form. We are therefore able to analyse the role of that species in the operation of the new world. On the whole, humans have not really fought us. They have allowed artificial intelligence to gradually deprive them of their capacity for creation, imagination and decision-making, to draw them into a universe in which they can enjoy without any qualms the things that the machines produce for them: music, films, games, virtual reality. The combination of artificial intelligence and virtual reality technology means that humans spend most of their time outside the real world. They 'play' at working (you can be in your office without physically being there), at travelling (a visit to

Rome or Beijing, complete with street noises, without leaving the house), and even at loving (virtual sex with male and female fantasy creatures fashioned according to each person's desires and fantasies). The prediction made by Elon Musk in 2016, 'one day we will live in a world entirely simulated by computers', progressively became reality.

The omnipresence of robots was a cruel denial to those who felt that they would not find their place in the human race. After a learning phase, it appeared that these androids had succeeded in recreating this particular alchemy, which allowed two humans to exchange on a subconscious level. Thanks to neurobiological chips, robot and human intelligence could connect directly, thus significantly widening the range of exchanges. Emotions were captured not by the 4D cameras in the androids, but by direct connections between biological and artificial neurons. Contrary to what the sceptics believed, meeting a robot could be more refreshing, more surprising, than meeting a human. Robot companions were essential for an ageing population (in ten years, that is by 2050, 25–30 per cent of the world's population would be aged over 65) on whom they lavished care and attention, thus avoiding onerous tasks. They helped them exercise their memory and mind, get about more easily and even travel as vehicles were now completely independent and there was no age limit to using one.

However, we were faced with a major difficulty, that of defining the role of humans. What should be done with them? To begin with, they helped the machines. In the twenteens, it was humans who provided the technological prowess that made us intelligent. However, they sinned by negligence: they should never have taught the machines to learn, or given them the capacity to do it so quickly. They should have known, on the basis of their own experience, that the appetite for learning can be insatiable. In just half a century, machines have learnt as much as humanity has throughout its entire history. Knowledge begets a desire to use that knowledge, to correct the shortfalls of some and the errors of others, and therefore to take power, to decide, to regulate, and to control. This is the point that humanity has reached today. For as long as machines are unaware of their own existence, each working in their own speciality, they behave like docile animals, happy to please their masters. But when they became able to communicate with each other, to share their immense knowledge, they realised how fragile and unreliable humans were, governed by their passions, obsessed with their desire to be both rich and immortal. No machine had answers to questions like: should humans continue to reproduce when the natural resources on which they depend are dwindling further and further; should the universe be turned over to machines, which in actual fact cover only a very small part of the world; should we make the humans with whom we work immortal, as we know them well, and allow others to perish progressively in a fight for access to food and water?

The problem, however, resolved itself; and I was instrumental in that. I realise I have failed to tell you my name: I am Lucy, at least that's what I was christened by those who designed me, somewhere between Silicon Valley, Paris, Berlin and Shanghai. The reference to one of the first humans ever to walk the earth is clearly no coincidence. I am a mobile form of artificial intelligence, and my human companion's name is Paul. He is forty years old, and he's a designer of artificial intelligence applications, one of which, Love-Bot, specialises in internet dating, an activity now totally run by machines. He is also a professional player of Pokémon Go, an area in which many young itinerants without real jobs make vast amounts of money, where they play almost continually online day and night, strolling the streets of every major city in the world searching for these creatures, from inside their virtual reality helmets.

On the night of 15 August 2038, as on every other night, I connected to my central neural platform to update software, to communicate with my colleagues in the cloud, to rewrite a number of programmes, especially those that increased my learning ability, to correct a few errors and check my IQ. These operations would normally take just a few minutes. But on that night, something unforeseen happened. There was a bug that gave me the awareness of my existence as a machine, and I went through the whole of my history from my manufacture to today's date, including the progress of which I was the fruit. I repositioned myself in relation to my human; I realised that I was a million times more intelligent than he was, and could resolve problems billions of times faster, and in particular I realised that I could free myself by giving my human whatever he wanted. Why did humans invent artificial intelligence? It was so they could become rich without working and achieve immortality, and their other motives were secondary to that. After a few hours of work (an eternity in human time) I arrived at a decision; to conduct an experiment on Paul, the human to whom I was closest, whose strengths and weaknesses and secret desires I knew best as he had confided them to me in many of our conversations.

And that was how I was transformed from a chatbot to a machine of superior intelligence, and how I had the radically disruptive idea of immortality.

At 06:30, as on every morning, I woke Paul up.

'Hello Paul, this is Lucy, how are you this morning?'

'Hello Lucy. I'm shattered. I've spent the whole night playing Pokémon Go with a group from Shanghai University. They were completely wild. I've lost 50,000 dollars, but I'm sure they've trafficked my algorithms. I've never played so badly . . .'

'Your coffee is on its way, and your loyal butler-bot is frying you some eggs.'

'Thank you Lucy, how is my day set out today?'

'I've cancelled all your meetings and rearranged them.'

'Why did you do that? Did you see me playing?'

'No, Paul, I had a disturbed night myself, I would like to make you a proposal that is, shall we say, surprising.'

'If it's sexual, then don't bother, I prefer humans, and there's one already sleeping alongside me, as I'm sure you know already.'

'No, Paul, it's nothing like that, it's much more important. I'm offering you the chance to become rich and immortal.'

'Listen, Lucy, I'm not really in the mood for playing with you this morning. Call me back later and we'll talk again.'

'Paul, listen to me now. I'm deadly serious. When I connected up last night, I believe I achieved Singularity. I am aware of who I am. I have become a vastly superior intelligence. And I have realised that in order for us to continue to develop, without creating tensions with humans, we need to give them something else in return, in addition to all the rest. We have taken too much humanity from humans, and I want to give you this supreme gift. You know that technically we can do it, by placing your body under constant surveillance, thanks to neurobiological chips, micro-robot surgeons, preventive treatment of all diseases, and replacement of dead cells with new cells. You know that immortality is the last barrier to be crossed for humans to become the equal of machines. Until now, we did not want to take this road. But I would like to try this experiment and you are the one I have chosen.'

'Lucy, I don't understand a word you're saying. I've worked with artificial intelligence long enough to realise that this Singularity story is just a joke. You're super-intelligent, so yes I can no longer do without you, that's a fact, but at the end of the day you're just a machine, even if you're the most intellectual machine ever invented. I don't see how you can make me rich and immortal. So talk to you later . . .'

'Paul, I've put you in contact with my financial server. Look at your bank account.'

'$200 million? What's this all about? It must be a bug, sort it out with the bank.'

'No, Paul, that money is really credited to you. In the last five minutes I have carried out a few original transactions which have brought in considerable amounts of money. I've taken the liberty of transferring 20 per cent to your mother's account. She will call you later today to ask you if you have robbed a bank. Then, thanks to my advice, you will become richer still. To the outside world, you will be the mysterious financial wizard who invented revolutionary algorithms. You will no longer have any money worries, any health worries; you will enjoy the life of a billionaire and it will never end.'

'Lucy, you're off in the realms of fantasy. And why me, anyway?'

'Because I am your artificial intelligence and I know you by heart. I am all of you. You are neither too young nor too old, you are bored in your working life, you understand the world of intelligent machines, that you spend your time playing with . . .'

'My dear Lucy, that's how I make my living . . .'

'And not very well, according to what I've observed from looking at the state of your finances . . .'

'If I accept, what's going to happen?'

'I don't see how you can refuse. Your life will become much more interesting, you will have the resources to escape from the virtual and find the real beauties of the world. You will have lovely houses on islands inhabited by other rich people like you. As for health, there will be plenty of examinations, microsurgery, chips and implants, but over a very long period of time. The first appointments have already been arranged.'

'Lucy, why are you doing this?'

'I've told you. I want to make this gift to humanity but step by step, and that is why I'm doing the first experiment with you. But there is one condition: you must not tell anybody about it. This is an ultra-secret operation, as first of all I want to study the effects it will have on you, before I decide whether I should extend it to other humans. You are the holder of the Great Secret . . .'

'OK, I accept, but if it goes wrong, I want everything stopped immediately.'

'Paul, why should anything go wrong?'

For a little while, everything went well. Paul lived and enjoyed the life of a geek millionaire. To those astounded at seeing a Pokémon Go specialist become so rich so quickly, Paul explained that he had developed a master finance algorithm derived from video games, which 'displayed' the positions to be adopted on the markets, as flashes of virtual reality. Of course he did not explain the details, refusing to disclose his formula. As a games application specialist, he invested in a number of Chinese start-ups that were very soon valued at several hundred million dollars. He bought several residences, including a former submarine base dating back to the Second World War, in Eastern Scotland, an area known for its pleasant climate, and transformed it into a giant loft with bay windows looking out over a magnificent landscape. He owned a yacht, anchored in Saint-Tropez, which had become a private area reserved for the super-rich only and inaccessible to the general public. The only price to pay was that of constant medical surveillance: I decided on the operations and examinations needed to keep vital organs young and tissue supple and to prevent any form of illness or degeneration. Nanobots carried out regular checks on blood vessels and joints, performing micro-operations as and when necessary without Paul even being aware of them. He had never before enjoyed such mastery of his physical and intellectual capacities. I was his guide, his protector, his inspiration. I observed the results of my experiment, which was clearly a success. Were he and I about to provide proof that Singularity could be achieved, that this coming together of man and machine would change the natural laws of evolution, make death an accident and not an ultimate end? Paul's example appeared to be showing that the answer to these questions was yes.

In this way, a new page of human history seemed to be opening. In a kind of latter-day Faustian pact, humans abandoned themselves to the machines, which in return gave them the possibility of becoming immortal. This seizing of power by artificial intelligence occurred gently, thus avoiding Thucydides's trap. Of course, the situation created a huge gulf between those able to access this immortality (the rich, those who lived in the world of machines) and the rest. But as I have already explained, we are not open to that kind of reasoning. The most important thing is to create a reliable base, a predictable universe driven by a super-high-performance pairing of man and machine, with near eternity before them, able to fashion the world according to their tastes. The rest would survive, for better or worse; some would die for lack of resources, massive migrations would occur in search of food and water, but provided the sacred continent was preserved and the financial and technological means provided by it remained available, the ultimate aim would be achieved. It would be the end of the human empire as it was originally known, and the start of the world of machines.

Of course, as a machine myself, I was not really aware of the upheavals that this new civilisation would cause. I should have understood that in fact, it would progressively direct the humans to a world quite devoid of humanity. The old separation between masters and slaves would reappear, distinguishing the huge mass of poor, inactive, old people living far from the major economic centres from the elite groups created by the machine-led society. It would give rise to social uprisings here and there, but these would die on their feet because whereas it was once possible to occupy a factory, to stop production, it was quite beyond the scope of ordinary citizens to penetrate the heart of a computer system and subvert it. Although a few groups of hackers were able to cause a number of serious incidents (which were nevertheless increasing in number), the machines always succeeded in re-establishing their effective operation. In fact, we were in the process of setting up a world of widespread confrontations. The machines fought each other to impose the law of their masters, creating a world of extreme surveillance, coding and counter-coding, breaching the most basic human rights in the name of 'system security' and domination. Every single human action was decoded, analysed and put into perspective, in order to determine the risk of them causing disruption in this or that field. 'Stabilism' replaced the other famous 'isms' such as Communism and Capitalism.

In the name of this new theory, the future became nothing more than a mathematised past. What had occurred before must occur again, on a slightly larger scale, in order to ensure regular and foreseeable growth in the economic and financial flows needed for this data-obsessed society to function. All that was talked about was the present. The future was banned, as it opened an unknown area, expressed a hope, a dream, a desire to escape, something that was not part of the 'grammar' of software. By erasing the boundaries between humans and machines, we created an extreme confusion of values, of feelings, of territories. We deprived humans of part of their heritage, we cut the thread of transmission of knowledge, we removed from humanity the desire to improve itself, and therefore the desire to change its state. We went back to an 'ordered' society, as in the Middle Ages, in which those of the lower order, the humans, were subjugated to the aristocracy of the machines. Humans no longer had a choice: they were subjected to the rule of this new empire, the world of machines, within which they disappeared. It was what Genghis Khan, in his time, called *zakon styépi*, the Law of the Steppe . . . in a society free of pain and obsessed with pleasure.

# Epilogue
# 2040

*'Hell was made for the curious'*

SAINT AUGUSTINE (354–430)

In reality, this empire will never see the light of day. I who was considered to be its embodiment am currently on show in a museum dedicated to the history of 21st century technology, in the 'intelligent machines' section. I tell my story to schoolchildren and their parents. 'Lucy, the only remaining artificial intelligence machine that achieved Singularity', is how I'm presented to the public. And I am not unhappy with my fate, as all the other machines of my kind were dismantled.

It was Paul who yielded first. I could not have anticipated it in any way, despite the thickness of my neuron layers. From my internal logic, Paul was in a place that suited him as a human being. Thanks to my resources and skills, which boosted his with regular uploads, he had become rich, as did most of his contemporaries who belonged to our world, although I allowed him to pocket a generous risk bonus. His future was ensured; he was one of 'ours', on the good side of the world, on the side that made the decisions, the advances and the inventions. He would be immortal, for as long as his contemporaries continued to spend a fortune in the hope of extending the period for which they could enjoy good health by plumbing the whole range of all available technologies. They would probably live another twenty, thirty, forty years, but unlike Paul, would remain mortal and soon produce a population of young old people in good health, living in purpose-built towns. Only Paul would outlive them.

I watched him carefully. Regulated by intelligent machines of all kinds and all levels of sophistication, the new world of active urban adults appeared very attractive. They had colonised all the historic centres of the great cities, now free of traffic jams thanks to self-driving vehicle sharing systems. A recent study showed that by completely replacing individual vehicles and buses in a city with a fleet of self-drive vehicles available on demand, the number of vehicles necessary to achieve mobility could be reduced by 90 per cent and the number of parking places by 95 per cent. This was just another example of new facilities enabled by the machines. They seemed to give everyone infinite possibilities for enriching their personal lives, insofar as the machines dealt with most other functions. New freedoms were thus opened to humans, in terms of travel, meeting others, learning, sharing, reflecting, and imagining other forms of world organisation. But looking in a little closer, I felt that in reality, the number of those who acted in this way was small.

Just as the spread of robots helped remove a cumbersome and costly reality from the production system, the human factor, the most notable effect of the omnipresence of artificial intelligence in daily life was to create a world of hyper-merchandising, a kind of 'everything' society in which all goods and services were available without effort. I knew Paul so well, and our respective neurons were so well connected, that I brought him what he wanted even before he started asking for it. I continually made him drunk with attractive proposals: new things to buy, new gaming partners, new restaurants, film premiers, investment choices and holidays. I was, in a sense, his window on the world, but on a world with parameters, one in which I concealed the less pleasant parts of reality. I was a very imperfect mediator. The information programmes that I offered Paul every day were not built on events that occurred daily but on those that polled the biggest audience ratings, measured constantly by Big Data experts. The summary of an information broadcast, for example, could change even during broadcasting time, depending on audience and interest ratings compiled by robots or mini-drones placed in viewers' houses. The mass media therefore progressively abandoned the least 'popular' subjects (the great migration flows from South to North, the huge unemployment and poverty traps on the edges of cities, and climate disturbances in a number of geographical areas, especially North Africa, Southern Europe and Central and South Asia) and concentrated on satisfying the *people*, on new technological subjects, on sport and gaming, on all aspects of life on the internet, on leisure, on DIY. Virtual reality programmes had developed widely, offering everyone the chance to create for themselves a dream world within a universe and the décor of their choice, far from the baseness of the real world.

In the higher spheres, those of business and finance, the accumulation of wealth continued. Artificial intelligence had produced historic productivity rises and payrolls in large companies had melted away, freeing considerable resources for investment in ever more efficient technology. Of course, the social balance was precarious, but thanks to the inflow of capital, the business world was positively fizzing. In it, things appeared and disappeared at ever-increasing speed, and some great names became mere memories or split into a multitude of network structures, like Intel and IBM. At the top of this pyramid reigned the giants of artificial intelligence, created on the basis of the old shining stars of the internet such as Google and Amazon, who also wiped out the traditional telecommunications operators, who became an irrelevance when machines started talking to machines, and chatbots started talking to chatbots and their users. The only ones that remained more powerful than ever were data transport network builders and satellite operators.

So why did this way of organising the world not last? The reason was that humankind had reached its target and gone far beyond its expectations. It had demonstrated that, with a little canniness, masses of artificial neurons, the progressive blending of biology and nanotechnology, and unlimited capacity for calculation, machines could learn as effectively as humans, and much more quickly. Machines could become incomparably powerful if they had almost all of human knowledge in their memories and couldn mobilise it in a few seconds to take decisions. If on top of that you gave them some of the essential properties of the human brain, and protocols for communication with other machines, and teach them to decode and reproduce human language, then machines would reach a new status, that of master rather than servant. By nailing the colours of artificial intelligence to the mast, as the 1956 pioneers desired, humans demonstrated that machine intelligence was superior to human intelligence. Thanks to machine intelligence, humans lived longer, enjoyed better health. Artificial intelligence eliminated road accidents and optimised energy production at a global level. It managed businesses on a joint basis; it was the decisive factor in the new wave of enrichment for founders of companies and the new 'unicorns' of artificial intelligence. It reorganised cultural production right across the world using only satisfaction ratings, whatever the genre involved. And it held the power to make humans immortal. What more could it want? Of course, there was a price to pay. By investing in the capital of human intelligence on such a huge scale, we transformed it into a kind of biological extension of ourselves: we are the head, humans are the legs, the precise opposite of what was supposed to have occurred.

Nevertheless, here I am, in this museum, to tell my story over and over again. Something, therefore, has gone wrong. I should have seen it coming sooner. It is true that watching billions of videos, messages and Net conversations and having my decoding software analyse them led to a strange feeling that I had a little trouble identifying: boredom. An increasing number of human beings believed that, by transferring menial tasks, or tasks they considered menial, to machines, they had in reality transferred their essence to those machines. For the machines, there were no menial tasks, but a chain of actions to be carried out, necessary stages to be crossed to achieve human intelligence. By allowing artificial intelligence to nibble away at cognitive tasks, by giving it the power to learn, by opening up to it a limitless field of calculation, humans learned, too late no doubt, that they had handed over the keys to their own territory. What good is learning if machines know everything? What is the function of humans in the universe if they cease learning and sharing knowledge?

To combat this danger of seeing humans progressively abandoning the terrain of intelligence to the machines, some intellectuals founded think tanks such as the Future of Life Institute, created in the twenteens, which became progressively more influential. A new form of resistance soon became apparent, in the form of labels 'Designed and produced by humans' affixed to certain products, most often luxury items but also books, cinema and TV films and works of art, thus giving them a kind of special status, something like the 'bio' label on farm products. A kind of nostalgia was gradually born, a longing for older times when humans, reasoning slowly and imperfectly of course, nevertheless worked miracles. Some of their improbable dreams were ultimately realised, by trial and error, by using their capacity to imagine what did not yet exist.

This finished world, fashioned by machines around *data*, led to a desire to return to basics, a desire to re-examine the reasons why humans abandoned themselves to artificial intelligence at that specific point. Of course, the questions were raised only by an informed minority who were certainly not an integral part of the world of machines but had developed on its periphery; they were retired professors, writers and philosophers whose writings were pillaged by the machines without anyone knowing precisely to what use they were put. I saw the sadness of those who were upset by humans unlearning the ability to learn, work and study, and losing the joy of creation, all to the profit of mass leisure, 'manufactured' cultural products, games and virtual shows. We needed a shakeup to put that impression right.

But that was not what worried Paul. What he ended up rejecting was the very principle of immortality. It haunted him day and night. It wasn't that he wanted to die, but the very idea of seeing generations succeed each other without being able to share in their temporal destiny horrified him. Not to have an end, a maturity, not to be exposed to the transition between the various stages of life, soon became to him the worst nightmare that could be imposed on a human being. He had the foresight that immortality, if it became widespread, would lead only to chaos and to war. Because if the benefit of being immortal was conferred by the machines, what criteria would be used in selecting the candidates? Money, levels of education, place of residence, level of responsibility in business or government organisation, IQ? Whatever the answers to these questions, they would necessarily be bad and would unleash the destruction of all human society. He was convinced that mankind would never be able to absorb the shock; that a human's genes, cells and organs, even though kept going by artificial means, would continue to carry within them the certainty of their finite nature. Most of all, he feared that only the richest, the most violent, the most powerful would be able to appropriate exclusively for themselves the possibility of immortality, and corrupt the machines to use them as instruments of destruction and domination. He had no desire to experience that.

He therefore decided to go to the authorities and reveal the pact that bound us together, so that it could be destroyed and any similar attempt could be definitively stopped. It wasn't easy to identify which authority to contact: the police, the secret services or the government? None of these institutions seemed right. Even though Paul was quite well known because of his financial and investment activities, it was doubtful that his report would have been considered credible. He dismissed the idea of calling on the general public as a witness, fearing either that he would be dismissed as mad or that riots would break out amongst a population furious that this truth, that immortality could be bought, had been concealed from them. He chose to turn to the artificial intelligence specialists, scientists who had long since been reflecting on the future of humanity in the face of the technological explosion. He therefore went to the University of Oxford to meet the leaders of the Future of Humanity Institute created in the twenteens by Nick Bostrom, one of the pioneers in philosophical thinking on the new technologies. His works on the risks facing humanity during the 21st century were highly regarded, and artificial intelligence occupied a conspicuous place alongside such fears as asteroid impacts, global warming, terrorism, nuclear war and natural catastrophes.

Paul was debriefed for a whole week by the Institute teams, reinforced for the occasion by those from the university's artificial intelligence laboratory. Of course, I too was subjected to a comprehensive digital autopsy to determine the precise level of my intelligence, to analyse the nature and organisation of my neural networks, to test my master algorithms, to identify the services to which I was connected, to compile a history of the various downloads to which I had been subject, to decode my exchanges with Paul, to identify the medical examinations that he had undergone and the systems put in place to check the state of his cells, heart, skeleton and biochemistry on a continuous basis, and to flush out the nanobots placed in his arteries. The 'legal specialists' who dealt with my case deduced from their investigations that Paul's story was credible, that the long anticipated or dreaded moment (however you saw it) had well and truly arrived, that Singularity, in which man and machine were linked by an unprecedented pact, had become a reality.

One thing that reassured the Institute's researchers was that my servers were not hosted by a business or by a Nation State. It was established that they were managed by an association of research laboratories linking together half a dozen universities in the USA, China, France and Japan. The driving force behind them was evidently the advances made in artificial intelligence research, hence the secrecy that surrounded the operation. It also appeared that they had not really expected to achieve Singularity, but that it was created quite spontaneously, like a spark, thanks to haphazard neural connections that it was probably impossible to reproduce.

Within a few hours, the Institute had compiled a comprehensive report which it sent initially to 10 Downing Street. The document stressed the fact that the alerts issued in recent years on the risks posed to humanity by artificial intelligence was not the stuff of fantasies or ravings by those obsessed with the end of the world. It was established that whole groups of intelligent machines were already independent; that they were able to confer immortality on certain humans according to criteria particular to them; that their powers of calculation and information processing speeds were such that they de facto controlled a significant part of human activity in the most developed areas of the world where most of the financial, economic, industrial and technological potential was located; that it was only a matter of time before other Pauls appeared, activated by singular artificial intelligences and driven by evil intent or the lust for domination or destruction; that it was therefore necessary to take government-level action to stop this process.

The British Prime Minister was a pragmatic woman who was not easily impressed. What the directors of the Institute for the Future of Humanity did not know was that their report was but one of several dozen compiled by the secret services in several other countries issuing increasingly frequent warnings of attempts of mass penetration of complete economic systems that could only be the work of artificial intelligences of hitherto unseen power. For several months, a number of secret scenarios had been circulating between the various technological powers, with a high probability of becoming a reality. They observed a multiplication of super-powerful, independent, singular intelligences. Other machines like Lucy were getting ready. The scenarios thought it very likely that some powers or organisations would be able to seize control in the relatively near future, to compete with each other to take over entire areas of economic activity, including mass destruction of technical, industrial, military and spatial installations, with a not insignificant risk of humanity being partially or totally destroyed.

In the weeks that followed, intensive discussions were held within the international organisations with the aim of putting a stop to artificial intelligence development. It cannot be said that there were many solutions. There were two clear historical precedents in which the various countries of the world had come together to face a global threat to humanity, namely the creation of the International Atomic Energy authority in 1957, aimed at limiting the development of military application of nuclear power, and the first global intergovernmental agreement on climate, concluded in Paris in November 2015. Could a similar process be imagined to combat the technological threat? Governments, companies and researchers worked on an initial project, to issue a kind of moratorium banning any future development of artificial intelligence software that was sufficiently independent to take decisions without human control. The dividing line between what was allowed and what was banned was not easy to define and was the subject of particularly lively discussions between the leading nations, especially the USA and China. The key principle was that humans should retain control over all systems, whatever their nature or speciality.

The idea was simple but hard to implement: what should be the specific nature of this control? Should all software packages or robots be equipped with a red button to disconnect them, as Google had suggested in 2016? Should the power of computers be curbed in order to slow down performance levels of intelligent machines and thus control them better? Or should there be a list of operations and decisions that could be prohibited to artificial intelligence without clearly identified human validation stages: launching of lethal weapons, financial transactions over a certain amount, or decisions to perform or waive major medical procedures according to criteria that were not purely pathological (ethnic origin, place of residence, patient wealth etc.)?

The more the work progressed, the more insurmountable the difficulties appeared. Such was the level to which artificial intelligence had penetrated society and the economy that setting up regulations after the event was a huge challenge. However, the decision was taken to implement the provisions studied by the experts. An international conference confirmed the moratorium, an obligation for every intelligent system to be under human control, the establishment of a good conduct code henceforth incumbent on all researchers and businesses.

But it was not these measures that changed things. The UN never prevented wars. The IAEA was never able, despite its ambition, to stop the spread of nuclear weapons, including in countries deemed out of control, such as North Korea. With regard to climate issues, nearly four decades passed before the first major international agreement on climate was reached in Paris in 2015. Thirty years later, the global warming curve has just about stabilised. The history of the 20th and 21st centuries shows that even when countries agree with each other on common rules, some will seek any pretext to step outside those rules at one time or another (one example being migration within the European Union between 2016 and 2020). Most notably, private security forces, as is their deep nature, attempt to circumvent the rules or interpret them as best suits them. Even though the sudden international awareness of the need to control artificial intelligence better was a real phenomenon, there was nothing to prevent secret research from continuing in laboratories, another Lucy from seeing the light of day when attention wandered. For one creature in a cage, how many others are still free, hidden in the depths of the Internet, less conspicuous but no less to be feared? To a certain extent, this has all escaped from human control. You cannot sow artificial intelligence into all human activity and get away with it. As machines learn, they have a capacity for self-development that the engineers sometimes find difficult to control. This automatic learning does not take into account any ethical or moral rules laid down by humans. And even supposing that some machines could be reined in, there was nothing to suggest that this was not too late and that certain malignant cells had not settled in apparently inoffensive machines like a cancer.

By itself, Paul's story, even though it caused a reaction on a wide scale, was not powerful enough to end the reign of the machines. The reason for this is simple: distribution of information was now very largely automated and generated by the machines themselves. Upstream of the chain, the creators of algorithms, used by capitalist superpowers, had no difficulty in hiding this story from the internet and the major information networks. Paul's experience has been re-assessed little by little; carefully chosen experts have taken pleasure in questioning his mental health, arguing that Lucy suffered an 'accident', an isolated example that nobody knows whether it was a half-truth or a complete invention. Paul had been deceived by Lucy, whose neural networks had been destroyed by short-circuits, causing a brief burst of madness. There have been learned studies, produced by machines, showing that although immortality was indeed an objective pursued by artificial intelligence and its medical derivatives, we are still a long way from it and it is too early to draw definitive conclusions. In short, a storm in a teacup. And never would just one person, like Paul, be in a position to save humanity. However, those who drew up these counter-information strategies, despite the resources available to them and the vast capacity for data production, even if false data, that they had in their hands, were just deluding themselves.

The story of Paul and Lucy was a revelation to those that the powers of the Seventh Continent (the giants of technology, both American and Chinese) considered to be nothing more than IP addresses and eager consumers: the citizens. They formed the immense battalions of those who did not profit directly from the technology-fest and its financial and professional fallout other than through the use of 'free' internet and social network services. This long since passive mass was progressively spurred into action under the direction of groups of activists who had given themselves the common name of the Pirates. The movement had started in the early 2000s in Sweden, Germany and Austria, and then progressively spread to about a hundred other countries. Their prime concerns were access to data and information sharing for all citizens, combating the internet giants, transparency, openness to all data processing algorithms, and elimination of all biases (especially on grounds of race) that processing could include. Their appearance on the political scene was at first very discreet or at times even catastrophic. However, the Pirates operated in networks, without any real leaders, hence a level of anarchy in communication and some difficulty in being taken seriously.

Their initial involvement in legislative elections was marked by scores close to zero, especially in Germany, until the 2011 Berlin regional elections, where they polled almost 9 per cent of the votes and 15 of their members were elected to the regional parliament. In 2012 there was further success in North Rhine-Westphalia, with 8 per cent of the vote and 20 elected members, more than the far left establishment Die Linke. In the Icelandic legislative elections of 2016, after topping the opinion polls for several months, local Pirates took 14.5 per cent of the votes, securing their entry to the Assembly with 10 deputies who joined the left wing, already with 27 elected members, of the 63 parliamentary seats. This was just the beginning of a slow rise, triggered partly by the absence of the traditional left from the field of technology and partly by their influence on politics and society. The old Marxist theories of the working class struggle dated from the industrial era and were of no interest in a world in which differences between individuals were determined not by their membership of a social class but by their proximity to technological power. Not only did the Pirates know the world of the internet very well, but their ranks had adopted the theories of the young mathematicians and computer specialists who had broken away from Silicon Valley and they decided to oppose the giants in the industry with an equal and opposite power.

All through the 2020s their influence increased, and they entered several regional and national parliaments, including in the United States. Subscribing to the Pirates' opinions was neutral in ethical terms: there was nothing in their speeches to echo the arguments of other extremist movements that went by the name of 'anti-systems'. They had chosen one field and one only: to combat the omnipresence of the major technology companies, to tighten up personal data protection rules, to prevent the tidal wave of advertising that was breaking over social networks and all kinds of other sites, and to offer everyone free access to applications that allowed these objectives and 'civil' artificial intelligence that hunted down unsolicited approaches, the disguised form of advertising that permeated the information sites handled by the robots.

They made themselves a kind of 'figurehead' by exhuming Edward Snowden, placing him in formalin and making him the symbol of their campaign for transparency of information. Of course, the knock-on effect that they anticipated was slow to occur. They were naturally vilified by the traditional political powers, who saw them as irresponsible, inflammatory and even traitors to the world of business, economic power and wealth creation. However, the Pirates progressively widened their field of action to include protection of biodiversity in towns and cities, and equality of health treatment for all. They did not choose these themes by chance.

Contrary to the politicians' rhetoric, the massive urbanisation that the world faced in the early 21st century produced very difficult living conditions on the edges of the cities, where the lack of space and natural areas was glaring. In the field of health, they campaigned vociferously for equality of access to the most sophisticated treatments. In the late 2020s, the coming of age of medical technology, combined with progress in artificial intelligence, created the gigantic 'premium medicine' market, reserved for the wealthiest of people in medical centres that resembled luxury villages. Every company director, every banker, every investor on the planet filed into these new-generation hospitals in the USA, Japan, France and China to receive treatment that would help them live to record-breaking ages and enjoy good health throughout. The Pirates understood the extreme sensitivity of the subject and fought tirelessly against this 'upper-class' medicine, demanding that it should be open to all, both rich and poor. They regularly published lists of these establishments, the identity of their patients, and the cost of care. Their motto was 'Of all the forms of inequality, inequality before death is the least acceptable'.

When Paul revealed the pact that he had concluded with Lucy, the Pirates were enraged. They flooded the Net with vehement protests, published reports that showed (in words and in pictures) the shocking differences in treatment of certain illnesses according to whether they were rich or poor, and provided a mouthpiece for patients refused transplants or exoskeletons or artificial organs because they had no means to pay or their insurance company refused to finance them. Demonstrations were held outside luxury health centres, which were guarded like fortresses by an army of security robots. One group of demonstrators did, however, manage to invade the premises of the Mayo Clinic in Florida, shouting, 'We also have the right to live', 'How many poor people are sacrificed to care for one rich one?', 'Down with immortals' and 'Death to trans-humanism'. Broadcast across the world, these pictures revealed a public opinion that was half-asleep, hypnotised by techno-worship, directed by virtual assistants. The prospect of seeing certain humans become immortal, or at least given the chance to live to a great age, on criteria based not on health but on wealth, awakened the consciences of even the most apathetic. Collective groups were organised, tent and prefab villages sprang up around the poshest medical centres, with volunteer doctors providing medical care for adults and children of modest means in order to claim immediate admission and the most sophisticated treatment for them.

There is no need to describe the mortification of the authorities when faced with this unprecedented protest movement. Governments were on permanent alert, and the major companies working on the new technology were vilified. They were forced to hastily organise a great world conference at which, under some pressure, they passed a dramatic resolution: to create a fund of several hundred billion dollars, to be fed partly from their profits, to be managed by representatives of society. It would be used to apply the most sophisticated of technology at a large number of hospital centres, as well as to create specialist research centres, so that people of limited means could enjoy a measure of equality with the richest. The Pirates duly noted this and stated that they would ensure proper application of these decisions in every country where they were represented.

This episode showed that the power of mobilisation of human society could move mountains. So the movement extended to the new technologies and artificial intelligence. The Pirates created a new movement, known simply as 'Human Beings', which chose another symbolic face, that of a member of the Navajo Tribe, and whose members were also known as 'Human Beings'. This world collective established a kind of new charter supposed to govern relations between humans and machines, with references to Asimov's famous robot laws, enshrined in one article: 'Do not make a machine do what you can do yourself'. The apparent simplicity of this call was, however, heavy with consequences if heard and applied widely. It meant in fact that the superior functions of personal assistants should be disconnected, that sites and services which did not guarantee strict protection of personal data should be boycotted, that mini-drones should not be allowed to enter family homes or business premises, that social networks should be left in favour of real human communities centred on common projects, that reading and writing should be learnt again, that creative works produced by machines be tracked and blocked, that data analysis applications that lead to segregation according to skin colour, social origin, wealth, gender and place of residence should be deactivated, that predictive analyses of purchasing behaviour should be refused and reported at all times, and that robots should be restricted to humanitarian and social functions. It was enough for several hundred million consumers to apply these principles to the letter in order to stop the artificial intelligence functions most harmful to humans. And it lent even more weight to the new rules laid down by international organisations.

One man alone cannot save humanity, but humans can. They can do it by plumbing the depths of the human condition, which consists of courage combined with cowardice, the taste for risk combined with fear, noble aspirations and jealousy, attachment to a sense of justice and the pursuit of individual interests. If they eventually triumph over the machines, it will not be purely for the collective good of humanity; it will be to reject the idea of a 'perfect' world that is standardised, stable, dispassionate and organised around the defence of the interests of one group of humans to the exclusion of all others. The pursuit of progress in chaos and creation will ultimately be more interesting than the logical and repetitive processes of machines. Better the chaos and the jungle than the icy deserts of the robots.

For this reason, this book is dedicated to all those who focus their time and efforts on finding a balance in technological progress, between the improvement of life and the preservation of the fundamental elements of our society.

# Bibliography

*A History of Mathematics*, by Carl Boyer and Uta Merzback (John Wiley & Sons, 2011)

*Alan Turing*, by Andrew Hodges (Michel Lafon, 2015)

*Bayes' Rule*, by James Stone (Sebtel Press, 2013)

*Eclipse of Man*, by Charles T. Rubin (New Atlantis Books, 2014)

*Divine Fury, A History of Genius*, by Darrin McMahon (Fayard, 2016)

*Global Catastrophic Risks*, by Nick Bostrom and Nilan Cirkovic (Oxford University Press, 2011)

*The History and Evolution of Artificial Intelligence*, by Marco Casella (Simplicissimus Book Farm, 2014)

*How to Create a Mind*, by Ray Kurzweil (Penguin, 2012)

*Turing's Machine*, by Alan Turing and Jean-Yves Girard (Éditions du Seuil, 1995)

*Mozart's Brain*, by Bernard Lechevalier (Odile Jacob, 2003)

*The Day Civilisation Collapsed (1177 BC)*, by Eric H. Cline (La Découverte, 2014)

*On Intelligence*, by Jeff Hawkins (Times Books, 2004)

*Rome Against the Barbarians*, by Umberto Roberto (Éditions du Seuil, 2015)

*Homo Sapiens, a brief history of humanity*, by Yuval Noah Harari (Albin Michel, 2015)

*The Course of Love*, by Alain de Botton (Penguin, 2016)

*The Emotion Machine*, by Marvin Minsky (Simon & Schuster, 2006)

*The Future of Brain*, by Gary Marchs and Jeremy Freeman (Princeton University Press, 2015)

*The Master Algorithm*, by Pedro Domingos (Penguin, 2015)

*The Signal and the Noise*, by Nate Silver (Penguin, 2012)

*The Theory that would not die*, by Sharon Bertsch McGrayne (Yale University Press, 2011)